THE NORTHERN NATURALIST

Experiences with wildlife in the Canadian Parkland . . .

Lone Pine Publishing

Copyright © 1983 by Lone Pine Media Productions Ltd.
First Printing 1983

The Publishers

Lone Pine Media Productions Ltd.
440, 10113 - 104 Street
Edmonton, Alberta

Printed and bound in Canada by
D.W. Friesen & Sons Ltd.
Altona, Manitoba

Typesetting and art by
Horizon Line Typecraft Ltd.
440, 10113 - 104 Street
Edmonton, Alberta

Edited by
Grant Kennedy
Cherise Vallet

Design by
Terri Guptell

Canadian Cataloguing in Publication Data

Hohn, Otto
    The Northern Naturalist

    Bibliography: p.
    Includes index.
    ISBN 0-919433-12-X

    1. Zoology - Alberta. 2. Zoology - Canada, Northern.*
    I. Title.
QL221.A4H63 1983   591.97123   C83-091433-1

43,429

THE
NORTHERN
NATURALIST

Written by:
Dr. E. Otto Hohn

*To My Sons
Howard and Peter*

Dr. E. Otto Hohn

E. Otto Hohn was born in Switzerland and obtained his medical and Ph.D degrees in England. He has been a professor of physiology specializing in hormones in animals at the University of Alberta since 1947. A bird watcher since boyhood, his interests were later extended to other animal forms. He is an elective member of the American Ornithologists' Union and, in virtue of field work in Arctic Canada and Alaska, a fellow of the Arctic Institute of North America. His papers on birds and mammals have appeared in *The Auk, Ibis, Arctic, Canadian Field Naturalist* and *The Blue Jay.* More popular items have appeared in the *Scientific American and Audubon Magazine,* as well as certain British and German animal magazines. Three short books, each on a particular group of birds, were published in Germany. On the local scene, he founded the Edmonton Bird Club in 1949 together with the late Professor W. Rowan.

Table of Contents

FOREST AND
WES

BRITISH
COLUMBIA

ALBER

PRINCE
GEORGE

EDMON

CALGARY

ASPEN PARKLAND
BOREAL FOREST
SUBALPINE FOREST
CARIBOO PARKLAND

RKLAND REGIONS IN
RN CANADA

SASKATCHEWAN

MANITOBA

ASKATOON

WINNIPEG

Photo Credits:

All photographs by Dr. E. Otto Hohn except as noted beside individual photographs throughout the text.

Assistance in the publication of this book was provided by Alberta Culture.

Sources

Banfield, A.W.F. (1974) *The Mammals of Canada.* University of Toronto Press, Toronto.

Bauer, K.M. and U.N. Glutz von Blotheim (1968) *Handbuch der Vögel Mitteleuropas,* Vol. 2. Akademische Verlagsgesellschaft, Frankfurt am Main.

Bent, A.C. (1937) *Life Histories of North American Birds of Prey,* Vol. 1. Smithsonian, Washington.

Bird, R.D. (1961) *Ecology of the Aspen Parkland of Western Canada.* Canada Department of Agriculture, Ottawa.

Dee Brown (1971) *Bury My Heart at Wounded Knee.* Holt, Rinehart and Winston, New York.

Dobie, J. Frank (1950) *The Voice of the Coyote.* Little Brown, Boston.

Graphic Services (1972) *Map of Forest and Parkland Regions in Western Canada.* Department of the Environment

Griffith, D. (1979) *Island Forest Year, Elk Island National Park.* University of Alberta Press, Edmonton.

Grzimek's *Animal Life Encyclopedia,* Vol. 12 (1975) Van Norstrand Reinhold, New York.

13

Hainard, R. (1961) *Mammifère's Sauvages d'Europe*, Vol. 1. Delachaux and Niestle, Neuchatel.

Holmgren, E.J. and P.M. Holmgren (1972) *Two Thousand Place-Names of Alberta*. Modern Press, Saskatoon.

Johns, J.E. (1964) *Testosterone-induced Nuptial Feathers in Phalaropes*. Condor 66: 449-455.

Johns, J.E. and E.W. Pfeiffer (1963) *Testosterone-induced Brood Patches of Phalarope Birds*. Science 140: 1225-1226.

Koski, W.B. (1981) *The Blue Jay*. 39: 197.

McDougall, J. (1971) *Parsons on the Plains* (edited by T. Bredin). Longmans, Don Mills, Ontario.

Murie, O. (1954) *A Field Guide to Animal Tracks*. Houghton Mifflin, Boston.

Nyland, E. (1970) *Miquelon Lake*. in Alberta Lands, Forests, Parks and Wildlife 13: 18-25 (now discontinued).

Palmer, R.S., Editor (1962) *Handbook of North American Birds*, Vols. 1 and 2. Yale University Press, New Haven.

Ryden, Hope (1975) *God's Dog (The Coyote)*. Coward, McCann and Geoghegan, New York.

Salt, W. Ray and Jim R. Salt (1976) *The Birds of Alberta*. Hurtig Publishers, Edmonton.

Soper, J. Dewey (1951) *The Mammals of Elk Island National Park*. Canadian Wildlife Service, Ottawa.

Soper, J. Dewey (1964) *The Mammals of Alberta*. Queen's Printer, Edmonton.

Stebbins, Robert C. (1966) *A Field Guide to Western Reptiles and Amphibians*. Houghton Mifflin, Boston.

Thompson, Seton E. (1912) *The Arctic Prairies*. Constable, London.

Young, Stanley P. and H.H.T. Jackson (1951) *The Clever Coyote*. Stackpole, Harrisburg.

Introduction

The Northern Naturalist is an informal guide to wildlife in the Canadian parkland. It does not pretend to be a comprehensive one.

For many years I have been enthusiastic about the varied scenery and wildlife of the Cooking Lake moraine country east of Edmonton. I have owned a small acreage in this area for 25 years. The birds and mammals to be found there occur throughout the whole parkland zone and though I write about experiences in the Cooking Lake Uplands (as I call the land over the moraine) these are but a sample of what may be met with over a much greater area. Not every species of bird or mammal of even this restricted area figures in my account for I have confined myself to those that seemed most interesting. On the other hand when dealing with some of the birds which only pass through this area on migration, I have written about my experiences with them on their breeding grounds on the Arctic Tundras of Canada and Alaska. Reflections based on the many years I spent in England also appear, as well as a glimpse of the South American wintering ground of one of our hawks.

In discussing the aspen parkland zone as a whole I've mentioned some examples of differences in animal distribution between a heavily wooded area like the Cooking Lake Uplands and the more open, more intensely cultivated surrounding area. Other instances are provided by horned and meadow larks as well as Swainson's hawks. These nest in areas around but not within the Uplands. In winter snowy owls are widely distributed but only once in many years have I seen one in the moraine country. Among small mammals Franklin's ground squirrel is characteristic of the uplands but avoids the open country while the situation is just the reverse in the case of the thirteen-lined ground squirrel.

In spite of some losses the present day aspen parkland fauna retains, as this book attempts to show, a great many interesting elements. I have dealt with animals and topics that seemed of particular interest to me without mentioning every species.

NOTE: In 1982 the American Ornithologist's Union published new "official" vernacular names for certain North American birds. Most readers, however, will be familiar with the older names which have been used in this book. To avoid ambiguity the new and old names of species mentioned in the text are tabulated here:

New Name	Old Name
Tundra Swan	Whistling Swan
Greater White-fronted Goose	White-fronted Goose
Northern Harrier	Marsh Hawk
Lesser Golden Plover	Golden Plover
Red-necked Phalarope	Northern Phalarope
Northern Hawk Owl	Hawk Owl
Northern Flicker	Yellow-shafted and Red-shafted Flicker
Yellow-rumped Warbler	Myrtle Warbler and Audubon's Warbler

I should also mention that where I have mentioned goshawk, kingfisher, sapsucker, magpie, crow, yellowthroat and meadowlark the full names are actually northern goshawk, billed kingfisher, yellow-bellied sapsucker, black-billed magpie, American crow and western meadowlark respectively.

PART I

The Aspen Parkland

The
Aspen
Parkland

A broad belt of aspen woods interrupted by patches of grassland extends for almost a thousand miles from southern Manitoba to the eastern foothills of the Rockies. It is the narrowest of the great vegetation zones of western Canada, forming a belt of about 200 miles from north to south in its widest stretches. It is, however, more picturesque than the gloomy coniferous forests which bound it to the north and the comparatively featureless prairie country to the south. Much of its charm lies in its diversity . . . the small and large groups of trees scattered about in areas of open grassland. This richly featured country is known as the aspen parkland.

Conversion to farmland has brought about the removal of the original tree cover in much of this area but there is still a great deal of undisturbed parkland in hilly country and in areas of poor soil. These factors occur together in land overlying old moraines and there are a number of these scattered across the parkland belt.

Moraine country is characterised by many small rolling hills up to 100 feet in height. In the low spots or potholes between these hills and hillocks there are many small or large sloughs and in the largest depressions there are a few shallow but sizeable lakes. Widely scattered boulders from the size of a coffee table to a granary attest to the former action of glacier ice which in its melting deposited these rocks after having carried them great distances. Little good soil overlies the glacial drift of the moraine. As a result the original poplar forest was not cleared as extensively as in the surrounding districts. The ridges of my home moraine have given rise to the name Beaver Hills. The area has also been called the Cooking Lake Highlands. However as these "highlands" lie only two hundred feet above the surrounding country, Cooking Lake Uplands would seem a more appropriate name.

The varied relief of the moraine country provides many interesting vistas on a moderate scale. Through a gap between two masses of woodland one may look down a steep incline at a slough with its cattails and marginal willows and beyond to rolling pastures bounded toward the horizon by fold upon fold of forested ridges. Occasionally one comes across a so-called boreal island where dense stands of young birches, often with some black spruce among them, grow on boggy ground. Probably there will be an undergrowth of labrador tea with occasional blueberries among its stems. These bogs carry outlayers of a northern vegetation typical of the boreal forests. The bogs are particularly attractive in winter when much of the country has a dreary aspect. The red branches of the birches and their white trunks add cheerful colour to the scene. Even the broad winding trails, now disused, which lead from the few roads that existed at the beginning of the century to individual farms, offer vistas of interest on an intimate scale. Though the trails themselves are generally still clear of trees, overhanging branches from the trees on either side form a nearly complete canopy above. The undergrowth among the grey poplar trunks of wild roses, hazelnuts, red osier dogwood and other shrubs adds to the variety. It is more readily appreciated while walking along one of these trails than when tramping through the "bush" where knees are inevitably scratched by thorns and where an unnoticed deadfall can often cause a fall.

The aspen parkland extends as a broad belt from west of the Edmonton area eastward to Manitoba. A further great patch lends variety to the Peace River Country. Though the aspen poplar is its commonest tree, there are also black poplars (also called black cotton wood) with deeply furrowed blackish lower trunks and balsam poplars. Other more scattered deciduous trees are the paper birch and in moist areas alders and willows of several species. There were originally considerable stands of white spruce but some were logged and many were destroyed by fires lit by the early settlers in their land clearing operations. Fire kills spruce completely but aspen can regenerate after fire by suckers from the roots. For this reason most of the stands of white spruce which remain are on islands in lakes where the trees were protected from fire.

The term parkland for the poplar zone is derived from the frequent small patches of woodland scattered about areas that were originally small prairies and are now generally pastures or tilled land. The resemblance to a park in the usual sense of the word is tenuous, but the poplar zone is certainly more park-like than the prairies to the south or the dense boreal forest which bounds it to the north.

Poplars can spread by sprouting suckers (rhizomes) from their roots. Where a meadow is bordered by poplar wood its edge gets invaded by poplar saplings. If nothing is done about this invasion a belt of woodland will in the course of years replace a sizeable strip of meadow and this process, if unimpeded, will continue with gradual shrinking of the grassland. I do not know how farmers deal with this problem. The method I adopt on our acreage is labourious. I try to pull up every sapling, often with a segment of root, or what is less effective, to cut it as low down to the ground as I can manage. Poplars of course also spread by wind borne seeds but these rarely develop when they fall on meadowland because of competition from the grasses. In unconscious support of the trees burrowing animals like pocket gophers, which are very common in the area, ground squirrels, foxes, badgers and coyotes produce heaps of bare soil. On these fairly small areas poplar seedlings have a good chance of survival.

The native Indians of the area, though they too set fires at times, had little effect on the country and its wildlife. It is likely that there were but few Indians in the Cooking Lake Uplands at any time. The missionary John McDougall mentions Indian camps, perhaps each consisting of only one family, at different points along the eastern and southern fringe of the Beaver Hills in 1865. Settlement by whites, on the other hand, has had gradually increasing and ultimately profound effects. Nearly all of them are unfortunate from a naturalist's point of view.

The advent of the fur traders naturally stimulated the trapping of fur bearing animals, at first mainly by the Indians. As a result,

Beaver beaver gradually decreased in numbers and by 1900 they were almost extinct in many areas. Protective measures and the restocking of former haunts has reversed this trend and the beaver population has for some decades been restored. In the Cooking Lake Uplands beaver lodges are now so common in large and small waters that it is difficult to imagine that the country could ever have supported more colonies of these animals.

Bison are generally thought of as roaming the prairies but they also lived in the aspen parkland zone and even beyond it as far

Bison
north as Great Slave Lake. By the mid-eighteenth century horses, originally introduced by the Spaniards to Mexico had spread to the Indians of southern Canada by means of trade, capture in raids and taming of animals running wild. With horses, buffalo could readily be run down on the North American great plains. When firearms were acquired, buffalo hunting became easier still. However the period of prosperity and relative comfort provided for the natives by the buffalo hunt lasted little more than a century. By 1870 bison were exterminated over virtually all their original range. The slaughter was mainly a result of the unchecked activities of the white hunters in the U.S. They were no doubt motivated by greed, but it seems that the extermination of the buffalo was also encouraged on political grounds. General P.H. Sheridan, distinguished for his part in the American civil war, was asked whether the wholesale slaughter of bison by white hunters should not be stopped. It is reported that he said "Let them kill, skin and sell all they can until the buffalo is exterminated as it is the only way to bring lasting peace and allow civilization to advance". (Hornaday 1889, cited from Dee Brown, 1971) The disappearance of the buffalo was in fact followed by the advance of white civilization, for most Indians not long afterwards found themselves confined to reserves in both the U.S. and in Canada. They were intended to become farmers but the fact that this and other programs for their welfare have not been altogether satisfactory is shown by the high suicide rate among Indians, by their disproportionate numbers in our jails and the prevalence of alcoholism on many reserves.

The grizzly bears and wolves of the great plains which used to follow and prey on the herds disappeared along with the buffalo.

No other result of the westward advance of the whites was as disastrous as the disappearance of the buffalo. Later on when other forms of wildlife were threatened protective measures were generally initiated before the situation became desperate.

In areas settled before 1900, big game for a while reached an all time low due to unregulated hunting by people who were at times near starvation as well as due to hunting regulations which were too liberal.

Farming during the first quarter of the century, when horses were still the principal sources of power, actually benefited some forms of wildlife. White-tailed deer which
Prairie Chicken
seems to prefer scattered bluffs of trees among fields instead of the uninterrupted woodlands that are the habitat of mule deer, increased considerably and appeared in areas

where they were previously unknown. The greater prairie chicken, a bird of grassy plains with scattered bluffs invaded the area, probably in response to the clearing of much forestland. It has since disappeared completely from Alberta and is extremely rare in other prairie provinces, perhaps as a result of changed agricultural practices. The sharp-tailed grouse (often confusingly called prairie chicken in rural Alberta) also became numerous during this period. Both these grouse found waste grain around stooks and threshing sites so that farming operations increased their food supply.

From about 1925 tractors began to replace horses and farming became generally mechanized. Tillage practices changed. The introduction of bulldozers in the mid and late forties stimulated the clearing of much wooded land. The increase of the rural population during this period was accompanied by a greatly extended road network. This and the more general use of motorcars made almost all the parkland accessible to hunters. While hunting pressure has increased with more disturbance of the game, there seems to have been little increase in the numbers of animals killed. This is said to be due to the greater proportion of unskilled hunters. It is certainly not uncommon to see would-be nimrods shooting at ducks hopelessly out of range. In view of the cost of shotgun shells the more parsimonious observer finds it difficult to understand such behaviour. Perhaps deer hunting skills too have deteriorated. Many hunters simply drive around ready to jump out and shoot at any deer they see rather than making the effort of a stalk or waiting patiently in silence at a meadow or along a game trail.

The whooping crane figures prominently among species lost to the parkland during the early part of the century. This crane, the

Whooping Crane tallest of all Canadian birds, is white with black wing tips and has a patch of red skin over the crown and below the eye. Formerly most of these cranes nested in the parkland. In the 1950's some were discovered nesting in small numbers in muskegs in the Southern portion of the boreal forest zone in Wood Buffalo National Park. One of the few recorded nesting places of these cranes in the parkland of Alberta was Whitford Lake, 25 miles northeast of the northern tip of the Cooking Lake Moraine. They are unlikely to have bred in the hilly and wooded uplands. I cannot believe that these large cranes were ever common, though one 19th century Albertan who lived on Buffalo Lake wrote of whooping cranes flying over "like a white cloud in the sky" (Nyland, 1970).

Whooping crane near Glasslyn, Saskatchewan

Nowadays the only remaining breeding area of these cranes is in Wood Buffalo Park in northern Alberta. The remnant crane population always winters on the Aransas Wildlife refuge on the Texas Gulf coast where the birds can be counted fairly easily. The hoped for increase in the number of these very rare birds can now be followed accurately. On their long migration between these two points whooping cranes regularly rest for some days in spring and fall on grain fields, generally near lakes, in central Saskatchewan.

While the whooping crane is still an endangered species the passenger pigeon, formerly a common bird of the parklands, is

Passenger Pigeon

totally extinct. This migrant nested in great numbers in Manitoba and Saskatchewan, it also occurred in Alberta. According to Salts' "Birds of Alberta" there is no proof that it ever nested in the province. However a nineteenth century observer cited by Edo Nyland (1970) reported that thousands of these pigeons nested around Pigeon Lake and gave the lake its name. The passenger pigeon was much like an enlarged mourning dove with a bluish head and reddish brown breast. It fed mainly on wild berries and after settlement was much attracted by field grains. It is reported to have existed in such vast numbers that flocks on migration darkened the sky and when they settled in trees would often break the branches due to the weight of their numbers. They were shot and trapped for food throughout their range at a rate which could not be balanced by their reproduction. Like most pigeons and doves they only laid one or at most two eggs each year. The result was that the last wild passenger pigeon in North America was shot in 1900.

Other changes in the animal population of the parklands were less drastic. Black bears have generally disappeared but are still **Black Bear** found in some numbers in Riding Mountain National Park where they are protected. A few also live on the heavily wooded hilly areas of the aspen zone but these are only occasionally seen. Accounts from the fur trading era suggest that the cougar was never prominent in the aspen parklands, but there are recent sightings from all the prairie provinces including the Cooking Lake Uplands. These reports most probably represent animals that have wandered from a distance rather than a small resident population.

Of fur bearing animals marten, fisher, wolverine and otters are only found on the fringes of the parklands at present. All indica-**Badger** tions are that they were never well established in this zone. The badger was originally common in prairie areas of the parkland. As the holes it digs in fields and pastures made it unpopular with farmers and as its fur at times commanded a fair price, its numbers suffered with settlement of the land. It has however persisted particularly in the more open portions of the parklands. Mink, muskrat and, as mentioned earlier, beaver as well as three species of weasels have all held their own in the area. This is also true of the coyote and the red fox.

The explorers and early fur traders of the parkland region found the elk or wapiti common throughout the region, second in abun-**Elk** dance among the big game animals only to the bison. They generally called the elk *red deer*, probably because that animal was and is the largest deer of their European home lands. They were not far from the mark. Most modern zoologists consider the red deer and the American elk to be members of the same species because they are linked by many transitional forms across northern Asia. There is, however, a curious difference; in the rut the European red deer stag raises its head and roars like a lion. This is far more impressive than our much larger elk's bugling, which is little more than a hoarse whistle. There are still small numbers of wild elk in the remote heavily wooded areas of the parkland including the Cooking Lake Uplands and there are considerable populations in the National Parks of the region.

Up to the mid-nineteenth century the pronghorn antelope, though essentially a prairie animal, also occurred on the natural

Pronghorn Antelope

larger grassland areas of the parkland. It is now restricted in our region to the short grass plains i.e. the prairie country of southwestern Saskatchewan and southeastern Alberta. Its disappearance from the parkland was perhaps not only due to settlement but also because in this northern part of their range the animals were more often exposed to severe winters than on the prairies.

Fish were abundant in the rivers and lakes of the parkland before it became settled. Unfortunately our information as to just what species were involved is sketchy as writers of the period like Alexander Henry often used Indian and French names spelled in their own individual way. Catfish, sturgeon and white fish were mentioned. However, pollution from city sewage, the dumping of oils and other chemicals into rivers, and in some localities over-fishing have now depleted the fish population of the region to a very marked degree.

Settlement of the Aspen zone was accompanied by the total extinction of the passenger pigeon, the disappearance within it of wild buffalo and marked reductions in the number of several other mammals and a few bird species. Yet some native animals have become more common as a result of human interference with the land.

White-tailed deer were originally only found south and southeast of our area. They were unknown in Manitoba until

White-tailed Deer

about 1870. Their northward and westward spread was due to the clearing of forestland. Since they favor rather open country with only scattered bluffs of poplars they have spread throughout the aspen parkland replacing the mule deer which was originally common there. As noted in a later chapter this process was still going on in the Cooking Lake Uplands during the last thirty years.

The white-tailed Jack rabbit, originally a prairie dweller, finds cultivated fields as suitable for bedding and raising its young as its

White-tailed Jack Rabbit

native grasslands. With the spread of agriculture across the parklands it was able to extend its range. However in heavily wooded areas like the Cooking Lake Uplands it is still absent.

A number of observers have noted that crows seem to spread and increase in numbers with agriculture.

The members of the swallow tribe have increased in numbers because they have been able to use man-made structures as alter-

Cliff Swallow natives to their relatively scarce natural nest sites. Cliff swallows originally nested on cliffs where an overhang protected their mud nests from rain. Cliffs are scarce in the parkland but the swallows can now nest beneath the eaves of buildings and under bridges. Bank swallows nested in burrows which they excavated in sand or clay banks along water courses. Man has provided them with additional banks in road cuts and in gravel pits.

Barn swallows originally nested in caves or on cliffs with mark-ed overhang, both features which are virtually absent in the area,

Barn Swallow so that it was still a rare bird in Manitoba in 1909. It now nests on rafters and walls in-side buildings as well as under deep eaves. Barn swallows can discover suitable nest sites even in remote areas. I found a pair nesting at a fishing lodge in the middle of the Caribou Mountains where the buildings are the only ones for something like 30 miles in all directions.

Tree swallows formerly nested in tree holes generally made by woodpeckers. The number of suitable holes was increased by the

Tree Swallow provision of cedar power poles in which the swallows were able to use deserted holes that had been made by flickers. Bird boxes also increased the number of nest sites available to them.

The purple martin used to depend on old woodpecker holes for nest sites but has readily accepted the multichambered houses erected for it by man.

Until 1911 magpies were unknown north of Red Deer river. They have since spread northward and are common in Alberta as

Magpie far as the Peace River district. In recent decades they have also invaded cities like Edmonton where they now show a higher population density than in the countryside. As they are still scarce in unsettled areas their spread has most likely been forwarded by the advance of agriculture.

A few wild animals have also been successfully introduced to our area. There is the gray or, as it is known locally, the

Hungarian Partridge Hungarian partridge. This European game bird was first released near Calgary in 1908-1909 where it soon increased in numbers and is now found in the central and southern portions of all three Prairie Provinces. Numbers peaked just before the last

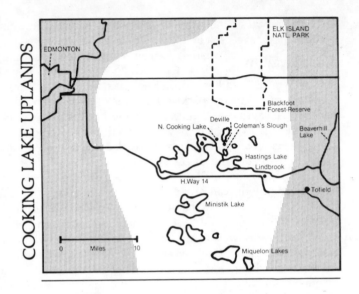

Map labels: ELK ISLAND NATL. PARK, EDMONTON, Blackfoot Forest-Reserve, Deville, N. Cooking Lake, Coleman's Slough, Beaverhill Lake, Hastings Lake, Lindbrook, H.Way 14, Tofield, Ministik Lake, Miquelon Lakes, 0 Miles 10

war and these birds are now more scarce. Pheasants were released near Calgary at the same time as the partridge. They too became well established but are no longer as common as they were thirty years ago. Both these game birds favour the cultivated country of fields with scattered trees and hedgerows.

Starlings introduced from Europe to New York in 1890 and 1891 gradually spread over most of the continent, reaching central Alberta about 1950. They have been com-

Starling mon for almost three decades and tend to live near habitations and cultivated land. As hole-nesters they compete with certain desireable native birds like bluebirds. Starlings make up for this in part by their varied if rather unmusical songs and the fact that some of them stay the winter when our bird life is so sharply reduced.

House sparrows, also of European origin, are even more common than starlings. They were liberated in several North American cities in the last century and,

House Sparrow although they are not migrants, have since spread over most of the continent. As their name suggests they generally nest on buildings but are not related to our natives like the song sparrow. A closer relative to the house sparrow would be the weaver finches of Africa and Asia.

PART II

Sketches of Some Animals

Sketches of Some Animals

Finding the Marsh Hawk's Nest

Every spring and summer we used to see two medium sized, long-winged hawks gliding about low over our pasture or perching for a rest on a fence pole. Occasionally one of them would suddenly lower its long legs, plunge down into the grass and almost disappear for some seconds as it snatched a mouse, grasshopper or even a small bird. These were marsh hawks, birds of open country but not, in spite of their name, restricted to wetlands. The males of this kind of hawk are pale grey with black wing tips, while the females are brown, so it was easy to tell that we were dealing with a pair.

When they came to hunt over the pasture they generally made their appearance over the southern skyline and they also disappeared in that direction. Evidently they were breeding somewhere south of our place and probably not very far away. Early one June I determined to try to find their nest. Beyond the ridge that formed our southern horizon there was a large, long-disused pasture with tussocks of wild grass and scattered shrubs. It was surrounded by poplar woods on three sides and from the western forest a tongue of young trees and bushes projected into the field. The nest might well have been anywhere in this open area and when I searched the male hawk did indeed show some anxiety. He circled about, every

31

now and then flying low over me uttering what I can only describe as chittering calls. However I found nothing and I also formed the impression that had I been close to the nest the hawk would have become more agitated than had been the case.

Another strategy seemed to be indicated. The male brings food to the sitting female from time to time. Marsh hawks and their relatives in other parts of the world do this in a special manner. They fly over the nest with prey in their talons and then let it drop. The female, which has been on the lookout for her mate, flies up and catches the falling morsel before it hits the ground. Other birds of prey feed the female tending eggs or young in the nest by bringing food to a nearby perch and giving a soft call. The female then flies over and takes her ration or, in some species, has it delivered to her on the nest.

It seemed that if I watched my male hawk from concealment I would sooner or later see him make a food-drop over the nest. I found an uncomfortable perch on the forest edge where the hawk evidently could not see me. He stopped his quartering over the ground and his calling and flew into a tall tree. From there he could no doubt look over the whole area that concerned him. I, on the other hand, could only see him and keep track of what he might be doing by leaning forward with great caution and watching through binoculars.

After a good rest, I thought concern for his mate would enter the hawk's awareness. Then he would start hunting and, if successful, take his prey to the nest. Evidently he felt otherwise or rather he felt nothing at all for he simply stayed on his perch. Whenever I painfully edged forward to look I saw him still immobile in his tree. This went on for an hour and a half until I left my cramped seat and, conceding the first two rounds to the hawk, gave up for that day.

Next morning I walked all over the "hawk's" pasture again, watching his behaviour carefully for signs of special concern. The hawk came closest in his down-swoops toward my head and called most persistently when I was in the tongue of trees and brush that projected from the western side of the field but there was no nest here either. When I stepped over the broken fence into the wood beyond he clearly became more alarmed. After pushing through a waist-high tangle of wild roses and some banks of nettles, I came into a fairly open area of rough grass with scattered young poplars and shrubs. The marsh hawk now made more daring overhead passes than before and I was evidently, at last, on the right track. Even so, I checked several of the clearings in vain until at last, as I entered a larger glade a pair of brown wings, looking huge at this close range, rose from the ground. It was the female marsh hawk flushing silently from the nest to vanish almost immediately among the trees. Her nest was little more than a heap of dead grass. It held four small white downy young with large black eyes. Beside them lay a bluish-white unhatched egg.

The nest had been so difficult to find because it was well inside the poplar wood, about 150 yards from its edge. Marsh hawks normally breed in the open, as I had expected this pair to do.

On my first return visit to the nest about two weeks later I had some difficulty in finding it again. When I did, I could only discover three young hawks. This time both parent hawks cruised overhead anxiously chittering but the female also uttered very plaintive "pee oo" calls.

When I checked the nest again after another week only one young hawk could be found and he was several yards from the nest. I was afraid a coyote or some other predator might have killed the others.

By mid-July the lone youngster, now fully 25 yards from the nest, was completely feathered. It could flutter along the ground but was unable to fly. This time only the female came overhead to display her anxiety at my presence.

At the end of the month it became evident that my fears for the fate of the young marsh hawks were unfounded. Four fully-fledged young were perched, each on his own fence post, not far from the nesting area. There they stood waiting for the parents to feed them. The old birds were surprisingly efficient at this. While I watched for half an hour, three of the young were each fed once. When a youngster spotted an approaching parent it set up a whining high-pitched call and, as the adult came still nearer, launched itself into the air toward it. The old bird then flew higher and dropped the prey it was carrying in its talons. The young hawk caught it as it fell and took it down to the ground to eat it. When, in a few minutes, it had gulped the food it flew back to its fence post and waited for more. Both parent hawks thus passed food to the young in the same way that the male delivers it to the female on the nest.

I had probably lost track of three of the young hawks until they reappeared when fully grown because they crawled even further away from the nest than the one youngster I had found on all my visits. Such behaviour would certainly increase the number of survivors from a brood. If all the young stayed at the nest during the weeks before they could fly, a predator finding them would kill the lot. Widely scattered as they apparently were, an enemy might perhaps discover one but not all of them.

Fully feathered young marsh hawks are coloured like adult females on the back and head but while she is pale brown below the young have rich reddish-brown underparts. For two weeks or so we were therefore able to see our marsh hawk family displaying three different plumages. But in mid-August, probably as soon as the young could hunt for themselves, they all disappeared from our vicinity although hawks of this species do not leave our area for their winter quarters in the States until sometime in October.

Next spring a pair of marsh hawks appeared once more and soon they were behaving as if they were nesting in much the same area as the year before. Yet this time I simply could not find the nest. However, late in May my wanderings in search of it yielded an unexpected bonus. In one of the open spots there was an unusual dead tree. Its trunk was broken off about six feet above the ground and the rest of the tree slanted down from the break with the crown resting on the ground. As I approached, a crow-sized brown bird suddenly flew up through the branches of this tree. It moved in absolute silence and in no time disappeared among the trees. However I had seen enough to know it was a long-eared owl. These owls usually lay their eggs in old nests of other birds in trees, but this one had simply laid a white, almost perfectly round egg on the ground, between some of the branches.

About a week later I found that the owl had built a simple nest of dead grass in the same spot and there were now six eggs in it. Now that it had a full clutch of eggs the owl made some show of defending them, using antics which might have intimidated some predators. On rising from the nest as I came on the scene, it flew into a nearby tree and there uttered a series of grating "quee, quee" calls while making itself look larger by holding out its wings alongside its body in threat gesture. Then it launched itself straight at me, passing within a foot or two of my head to land in another tree. Having seen the nest and its contents I left the anxious bird to settle down.

I was thrilled at this find for previously I had only once seen a long-eared owl in Alberta. Not merely to see another but to find it nesting close to our cottage was indeed satisfying.

Unfortunately this particular owl deserted its nest or met with some fatal accident. On all subsequent checks the owl was not to be seen and the eggs were always quite cold.

On the other hand the marsh hawks had fared quite well. In some spot I had been unable to find they had reared three young.

At the end of July, when they were already able to fly, they appeared on the same fence that the previous generation had used when waiting for their parents to bring food to them.

That summer was however the last during which there were marsh hawks near our place. A wide new highway was built across the rough pasture which had been their main hunting ground. Evidently this made the area unsuitable for them and they never came back. Neither did the long-eared owls, probably not because of loss of much of the pasture but because the road construction came too close to their nesting area. However a few years later in August one of these owls stayed in a poplar wood on our place for a while, suggesting that a pair had nested not too far away.

Moose and Deer

The advance of agriculture across the parkland zone has caused moose to disappear from much of their former range. Its strongholds are now largely restricted to its northern fringe. Yet it still survives further South in well-wooded hilly country. The Cooking Lake Uplands is such an area. In spite of the nearness of a great city and a fairly dense sprinkling of small farms and acreages, there is still a good population of moose. There are also white-tailed and mule deer as well as some elk in the district.

At first I found it difficult to believe that the elk and moose in such an area could be truly wild. I imagined they must be escapees from Elk Island National Park or the descendants of such escapees. However, after some enquires from the park superintendent's office I have changed my mind. I was told no bison had ever escaped from the park and that the park fences would also stop moose and elk from breaking out. Elk Island Park was set aside in 1906 to preserve a remnant population of elk that roamed the Cooking Lake Uplands. It is unlikely that every single elk in the area was manoeuvered into what became the park enclosure. The elk still present outside the park therefore in all probability represent what is left of the original wild population.

I have not had the good fortune to see an elk in our area but our district game warden assures me that they occur in some numbers in the Blackfoot Forest reserve. Elk have also been seen by some of the staff of Miquelon Lake Provincial Park as well as by one or two of my neighbours.

All deer, including the moose, lead very secretive lives hiding in the bush during the day and only coming out in the open to feed in the early morning hours and again towards dusk. These are certainly the times one most often comes across them but I have occasionally seen both white-tailed and mule deer grazing in a rough pasture near our place in midafternoon.

I think moose lead the most hidden lives of all. I had seen deer in the district for many years before I saw my first moose there, although I knew that others occasionally saw one and I was always on the lookout for them. So it came as a complete suprise when early one September morning I saw a bull moose across from me on the other side of a woodland slough. It must have been a fairly young animal for though it was full grown its antlers were rather small.

After that sighting my encounters with moose became more frequent. A year or two after the first episode as I was driving away from Miquelon Lake Provincial Park, again on a September morning, a magnificent bull moose suddenly appeared out of nowhere and crossed the track right in front of the car. This one carried an impressive set of antlers. It trotted along the lake shore at a good clip, its bell swinging from side to side, until a quarter of a mile away it turned into the bush and disappeared.

Another year a cow moose and her calf spent the winter in the immediate area of our acreage. I would come across their tracks and piles of droppings in the snow and find willows stripped of bark, far higher up than any deer could reach. We do not visit our cottage very often in winter but our country neighbours who live there all year saw the animals several times, on one occasion right in their front garden. It was not until the end of April that I came across them myself. During my usual walk along the boundary of our property I suddenly became aware of a cow moose standing at some little distance inside the poplar wood beyond our fence. She noticed me at just about the same time and vanished almost immediately among the trees. I heard a twang as she jumped the barbed wire fence which separates the woodlot from a secondary road. I thought the moose would cross this road and make for the bush beyond. Following as fast as I could I reached the road in time to glimpse the cow and her youngster, already beyond the fence on the far side of the road, as they were entering the thick cover there.

This year in May a friend was riding her horse through a wood near the north shore of Hastings Lake. It was a perfectly calm sunny day yet she saw what looked like a leaf moving in the breeze above a shrub. Taking her horse up closer she found that she had been watching the ear of a moose. It was a cow, lying on its side in the act of giving birth. The calf had already emerged beyond its middle. Alarmed by the presence of the horse and rider the cow clumsily got up and with the calf still attached withdrew further into the bush. Our friend did not want to disturb the animal further and left her in peace. No doubt the delivery of the new born calf was completed soon after that.

We ourselves have seen more of moose this year than ever before. One spring day our two sons and their wives were playing frisby when a yearling moose wandered out of the bush onto the adjoining pasture. My sons tried to head it off from the bush to which it was about to return so that we in the house might have a chance of seeing it too. At the same time the daughters-in-law raced for the cottage to alert us. The young men could not keep the animal in the open for long but we did catch a glimpse of it moving, at the nearest thing to a gallop I have ever seen in a moose, toward the wood it had come from.

In late November we had a real moose day. The weather was mild but foggy as we were driving on the highway to our place. Suddenly we became aware of a compact mass of black animals moving at a smart clip over the field beside the highway toward the poplar bush beyond. There was a cow moose, its striking hump towering above a sizeable youngster on its left and two others on its right. Unless one believes that a cow moose might adopt an orphaned youngster or two, this must have been a case of a cow with three young. Normally moose give birth to one or two young but triplets, though rare, do occur.

Later that morning I was walking along the willow-grown shore of a slough where I often go duck hunting when I was startled by a cow moose breaking cover ahead of me. As such close quarters her size was amazing. She trotted off some little distance, stopped, and turned to look at me for some seconds before sneaking off into the nearby wood.

I returned to this place a few days later and found her lying dead at the foot of the willows. One side of her body was leaning against the small trees and the legs were neatly folded under the body. Magpies had already pecked away the soft tissue of the lips. Otherwise what I could see of the body was intact. There is no open season on moose in this part of Alberta but some are nevertheless shot from time to time. I believe that is what happened to this animal, that it had been able to move some distance after the shot before collapsing and that the poacher had been unable to find it. To heave the animal over so as to look for a wound on the side that was hidden from me was beyond my strength so I was unable to prove my suspicion about the cause of its death.

During the cold months moose are often infested by a blood-sucking parasite, the winter tick. Miss V. Glines, a graduate student at the University of Alberta, has recently studied the effect of these ticks on their hosts. The ticks cause skin irritation which the moose deals with by grooming and rubbing itself against tree trunks or shrubs. A great many ticks are scraped off in this manner but so are a great many hairs. Severe hair loss results in moose which have a ghost-like whitish appearance. The white basal portions of the hairs, normally hidden by their blackish outer parts, are now exposed. Such moose are weakened, probably because loss of hair reduces the insulating value of their skins. In winter they may be unable to increase their food intake to balance their heat loss.

Miss Glines found that individual moose which removed fewer parasites by grooming, and therefore lost relatively little hair, suffered from anaemia.

The last time I cáme across a moose was in mid-April while driving along a gravelled road in a forested section. I saw a yearling male a few yards off the road at the edge of a still frozen slough. It was pitifully thin, its rump seeming to consist of little but skin stretched over the bones beneath. I stopped the car to watch and to my surprise the animal walked out onto the road and lay down there, making itself comfortable but keeping its head raised. I drove forward very slowly on the opposite side of the road but it showed no sign of being aware of the car until I was about ten yards away. Only then did it get up and unhurriedly walk back to stand where it had originally been. Its normal brown coat suggested that it was not tick infested or perhaps it was one of the animals which make little attempt to rub off the ticks. Yet its leanness certainly indicated that it must have had a hard winter. The snowfall had been heavier than usual and might have hindered its movements in search of food. It was clearly alone and the loss of its mother, depriving it of her guidance to good feeding areas and cover, may also have worked against it. Since spring was now well on the way this orphan would no doubt soon make up for past hardships.

In winter moose in our uplands feed mainly on willows and red osier dogwood bushes. Though in other areas they also browse on poplars they do not seem to touch them here. Birches are another winter food of the species, but there are not enough of them hereabouts to be a significant food source. Moose strip the bark off the willows, often to a height which they can only reach by getting the stems between their forelegs and riding them down to bring the upper parts within their reach. Willow bark, reached only by some exertion, and the tips of dogwood shrubs cannot contain many calories and moose must consume a great deal of this material during the short winter days to meet their food requirement. On and around our acreage, where in the last few years two moose have wintered, just about every dogwood bush has been worked over. The willows have been attacked too but in a more haphazard manner.

In summer moose live more luxuriously, not only on the leaves of the plants they use in winter but on a variety of herbs and grasses. At this season they also feed on water plants, wading,

Moose sparring

often with head submerged, in search of succulent plants, a habit which is not shared by any of our other ungulates.

I am reminded of two amusing experiences my friend Fred Rourke has had with moose. He lives on the shore of Hastings Lake and sees a great deal of moose. Sometimes one will come and scratch its skin on the water gauge Fred has inserted in the shallows below his house to record the lake level.

Last fall when these animals were in their rutting season, when males are known to be dangerous, Fred was in a wide open pasture with no sheltering trees in sight. He saw a bull moose emerge from the trees beyond the pasture making straight for him. At first he took little notice, as he felt sure the moose would turn off on seeing him. But it just kept coming on at a steady trot. A moose's trot moves him along quite fast, and soon the animal was pretty near, still aiming straight for Fred, who could not outrun it and had no shelter in sight. Only when it was about 40 yards off did Fred notice that a dog was following the moose, which was evidently trying to leave the dog behind and had no interest in Fred at all. At about 30 yards, the moose, never seeming to notice Fred, turned aside and passed him by as if it had never seen him.

In another moose episode resulting from Fred's somewhat impaired vision, he talked to a moose. Unlike St. Francis, he got no reply from the animal. The wife of a young ranch foreman, both good friends of Fred, often rides a dark brown horse over

Mule deer

woodland trails which are also moose trails in the district. As Fred was ambling along one of these, he noticed a large dark shape at the side of the trail. Obviously Julie's horse, the saddle girth of which she was adjusting on the offside, he thought. He called out a greeting. No reply. When a second remark was unaccountably left unanswered, he took a closer look and now saw he'd been addressing not Julie, but a very calm cow moose.

White-tailed and mule deer have shown interesting changes of population in the course of this century, not only in the Hastings Lake district but over much of Alberta. The early settlers found only the mule deer which was then fairly common. An account of the mammals of Elk Island National Park published as recently as 1951 only lists mule deer. The writer considered the white-tailed deer, which by then were found outside the park, to be recent arrivals in the area. Since that time they have managed to get entry into the park, the wardens believe by crawling under the fence, and have multiplied there. Nowadays there are about nine times as many white-tailed as there are mule deer in Elk Island.

Up to the late fifties I would come across mule deer as often as I would see white-tails in the Hastings Lake country. In recent years I see virtually nothing but white-tailed deer and they are evidently common. Two factors seem to have been involved in the change. Firstly, white-tailed deer have spread northward from their original range. Secondly, they are much more tolerant than mule deer to

41

the proximity of man and to the changes in the landscape brought about by farming. However, mule deer have not disappeared altogether from our area. Across the highway from our property is a rough pasture where in spring and summer we occasionally see white-tailed deer grazing. Sometimes it is a doe, at other times a buck with its antlers in velvet. It surprised us that they never took any notice whatsoever of cars passing at high speed little more than a hundred yards away. Early one morning last May, instead of one of the usual white-tails, there was a mature doe mule deer in this pasture. It is not difficult to tell the two animals apart. When white-tails move off they erect their tails which are surpisingly long, white on the tip and underside. They often wave them from side to side. Mule deer in my experience keep their tails down even when running away and in any case their tails have a distinct black tip. There are other differences. Mule deer have larger ears, the feature which accounts for their name, and the antlers of the two species also help to separate them.

Over the years white-tailed deer have provided me with a number of memorable sights such as a buck sailing with effortless grace over a roadside fence or, on a June morning, a dappled fawn only a few days old bounding up at my feet from some leafy undergrowth and then disappearing in a scrambling rush into the nearest tall cover. I saw my most impressive buck early one bright September morning. I was following my usual round of our land through a belt of poplars when suddenly I became aware of him standing between the tree trunks a scant fifteen yards away, partly in the sunshine and partly in the shadow. For some seconds he remained motionless but this could not last. He soon bounded into the bushes and through them into the adjacent pasture as I could tell from hearing the horses breaking into a gallop when his sudden appearance startled them.

These and other encounters with deer were always a pleasure but I once had a very unpleasant experience with a buck white-tailed deer. Admittedly he was in captivity and I am sure that what took place could never have happened in the wild.

On an overcast day in March I entered an enclosure which held several deer on Al Oeming's game farm. Within this enclosure was a pen with some ptarmigan. These were what I had come to see for I was doing some research on these birds at the time. As I walked toward the pen I noticed that one of the buck deer was looking at me strangely, showing a little of the white of his eyes. As it was long past the rutting season, a time when deer accustomed to people might become obstreperous and as I had been among these deer

White-tailed deer
Courtesy of Alberta Fish and Wildlife

with one of the farm workers only the day before, I decided to take no notice of the odd look of this animal and I walked past. Suddenly a violent blow from behind knocked me off my feet. In a flash I was lying on my back and the deer which had struck me from behind (above the knees as I could see later from my lacerations) now charged me from the front with lowered antlers. Grabbing an antler with each hand I tried to push him off but he could not be budged. He was straining downwards with the strength of all four legs behind him while I was pushing upwards. He pressed hard enough to shove me over the snow for twenty yards until my head came up against the ptarmigan pen. In the process I lost a shoe and my glasses. Of course I soon began to shout for help but with little hope. I had not seen anyone about for quite some time. Using my handhold on the animal's antlers I tried to twist his head but that did not bother him in the least. Then I held on to one antler and punched him in the neck with my other hand. This too was without effect. I am sure the buck would have enjoyed keeping me pinned down indefinitely perhaps until my grip on his antlers weakened when he could have jabbed me with his tines. Fortunately a chance visitor heard my cries for help and, picking up a stick, came toward us. As soon as he came within a few yards the deer backed off and trotted away as if nothing had happened.

I was not sorry to hear later that the antlers of this particular truculent stag were sawed off. It seemed the least he deserved.

43

The Coyote

A wild chorus of yelps merging into a series of shrill, high-pitched howls means coyotes. Their howling is most often heard at dusk or near dawn, more rarely during the night. I use the word 'chorus', for it sounds as if a whole group of animals were performing, yet in all likelihood no more than two or three are involved. In country with abundant cover like ours, coyotes are heard almost as often as they are seen. Their calls sound wild and carefree and suggest an exuberant enjoyment of life. This is in striking contrast to the long-drawn, deeper-pitched howling of a wolf, a sound that conveys a feeling of infinite sadness to the listener.

Coyotes are handsome animals, much like a smaller edition of a wolf but with a more slender muzzle and proportionately larger and longer ears. Being so much smaller, they look more dainty (I am tempted to say more elegant) than their larger relatives. Their colour is less variable than that of wolves. Grey above with a white throat and belly, they are rust coloured on the outer sides of the ears, feet and legs, and have a dark cross over the shoulders and a black-tipped, bushy tail. Canadian coyotes weight about 29 pounds or 14 kilos; wolves weigh at least twice that much, sometimes more than 100 pounds

When a coyote is seen well out on the ice of a lake, where its size is difficult to judge, it can be told from a wolf by the way it holds its tail when trotting or running. At such times a coyote allows its tail to hang, while a wolf keeps his up, in line with its back.

Capable of running at speeds up to 45 miles per hour when pursued by a car, coyotes are also excellent swimmers. One particular animal would repeatedly swim across the Columbia River at a point where it is nearly half a mile wide to get from its daytime haunts to certain poultry roosts on the other side.

In response to very heavy persecution by man, the coyote has historically extended its range over vast new areas. Originally, it lived from the Canadian prairies through the American west to central Mexico. The Aztecs of Mexico called it 'coyotl', and the Spanish conquerors of that country modified the name to the form now in general use. During the century following the conquest, coyotes spread southward as far as Costa Rica due, it is believed, to the introduction of livestock and land clearing by the Spaniards. Between 1830 and 1930, coyotes extended their range through the Yukon into western Alaska as far north as Point Barrow, and over the MacKenzie district up to the MacKenzie Delta. About the same time, they spread eastward, reaching central and southern Ontario, parts of Quebec, New England and New York state. The causes of their northern and northeastern expansions are not known. Perhaps persecution by trapping and poison in the original home of the species caused many individuals to wander further than was customary and thus to discover that they could find a living in new habitats.

The coyotes' success is in part due to its suspicious nature. If you see one from a car it is likely to remain unconcerned as long as the car is in motion, but it will take off as soon as you stop. It is equally suspicious of traps. Coyotes are intelligent, but the reader must decide for himself whether the anecdotes of clever coyotes which follow, plausible though they may be, are true.

A pet coyote was kept in a farmyard on a 30 foot long chain. The farmer's chickens soon learned to avoid the area the coyote could reach. One day when the little wolf was fed, it deposited some of its food a few feet short of the end of its chain, lay down and seemed to go to sleep. The chickens stared at the food, moved hesitantly toward and finally right up to it, at which time the coyote jumped up and grabbed one of them.

In a similar episode in Mexico, in an area where poison had been put out for coyotes, the observer saw a coyote suddenly fall down as if dead and lie motionless. A turkey vulture became interested, hopped up to the coyote and was seized and eaten.

The next episode is reported from a ranch house in Nebraska. The window sashes of this house were flush with the interior walls and the thick walls left a deep embrasure outside the windows. On a cold winter night, a cowboy in the house noticed a coyote asleep on the window sill, nicely sheltered from the wind. Stealthily, he raised the window, which was hinged at the top, and pitched his bull terrier onto the coyote. Then he closed the window and ran outside, expecting to see his dog fighting the coyote. But all he found was the dog, barking wildly, running round and round the house. Getting tired of this, he returned indoors and checked whether he had fastened the window securely. There was the coyote lying up against the glass as before. Evidently it had run once around the house and leaped back onto the window sill. Again the cowboy released his dog through the same window, and again the coyote returned to his refuge while the dog raced around the house. This time the cowboy decided that the coyote, by his superior intelligence, had won the right to his place on the window sill and left him in peace.

However, there are also instances of coyote behaviour which seem to indicate a lack of intelligence. The carcass of a horse left in the open was poisoned. Next morning, fourteen dead coyotes were found within sight of it, and the morning after that there were five more. Most of these coyotes must have seen some of the dead of their kind on their way to the horse, and yet this did not make them suspicious. This is perhaps not altogether surprising, since they are carrion eaters and will, on occasion, feed on carcasses of animals of their own species. However some of them must have seen one or more of their poisoned fellows in convulsions or showing other abnormalities of behaviour. The large number that came to the bait shows that even this did not put them off.

Another instance shows coyotes as less clever than crows. A farmer whose ripe watermelons were often tapped by crows injected some of the melons, in which the crows had already made holes, with poison. Soon there were some distressed crows about. Their fellows fluttered over them in great agitation, but from that time on never touched a melon on which there was even a barely started hole. Later, coyotes raided this farmer's cantaloupes. He injected some of the small ones with poison and placed them at the spot where the coyotes passed through his garden fence. Night after night, coyotes ate the poisoned cantaloupes and died.

Part of the coyote's success in the face of adversity may be due to the fact that it will eat anything capable of being digested. It even chews shoe leather or horse lines made of leather, perhaps for the mere joy of chewing rather than for the sake of the trace of oil it finds there. In our district, coyotes feed on rodents ranging from mice and voles to ground squirrels and true squirrels, as well as showshoe hares, carrion, occasionally deer, ground nesting birds, large insects like grasshoppers and beetles, and small smounts of such vegetable food as wild berries.

Winter must be the hardest time of the year for our coyotes, as for most animals that do not hibernate. An observation made by my friend, Dick Dekker, shows that at least some of them are well-fed, even in December. Only a 4 foot wide hole was left in the ice of Hastings Lake and a coyote was standing nearby. A mallard flew in and landed on the ice. It was obviously very weak, probably from starvation, for the coyote caught it with ease and killed it. However, it made no attempt to eat the duck, but left it lying and walked away. A second coyote then appeared. It grabbed the dead duck and playfully shook it about, but also left without eating any of it.

Coyotes are most successful at hunting deer when the snow is crusted so that it will bear the relatively light-weight predators but breaks under the hooves of the deer. Two coyotes will take turns to run the deer until it is exhausted. They have also been observed driving a big-game animal toward a member of their group hidden at some point on the route the hunted animal is likely to take. This would seem to imply that coyotes can learn the significance of game trials. A couple of coyotes may succeed in killing a deer fawn, though they may be no match for the doe. One of the predators makes feinting attacks at the doe, to which she responds by rushing after it. Meanwhile, the other coyote kills the fawn. A single coyote may sometimes, however, be defeated by a deer. A buck was witnessed chasing a coyote to exhaustion, then jumping on it with its forefeet. The coyote eventually managed to crawl into a patch of dense bush, where the observer found it in such a state that it could hardly crawl, so he put it out of its misery.

Further south, coyotes prey on antelope by much the same techniques as they use on deer. In this relationship also the predator on rare occasions becomes the victim. An employee of the Texas Game Commission saw eight fully grown antelopes chase a coyote for 8 miles. It finally took refuge in a bush. The antelopes followed it and jumped on it with their hooves. Finally a buck hooked the dead coyote out of the bush with its horns.

Some believe that the full grown deer and antelope which fall victim to coyotes are often sick or disabled (this is often a feature of predator-prey relationships). If this is so, coyotes might, by weeding out the unfit, actually have a beneficial effect on deer and antelope populations.

Compared to our own coyotes, those living further south have a positively exotic diet. In Texas they eat the fruit of certain cacti and on the coast, sand crabs. In some places they dig out and feast on the eggs of sea turtles. In Mexico the flower stalk of the agave (century plant) is cut out when the plant is ten years old. In the hole left by this operation, a sweet sap gathers which is collected daily. Allowed to ferment, it becomes pulque, Mexico's cheapest alcoholic drink, and repeated distillation turns it into tequila. Coyotes are extremely fond of the sweet agave juice, so that the men whose job it is to draw it off have to cover the holes with slabs of rock or other materials. In the southern parts of their range, coyotes also show an interest in such items as peanuts, carrots, tomatoes, peaches, apricots and plums. In desert areas, they know how to dig for water and make holes 2 to 3 feet deep to reach it in the sandy bed of dried out water courses. The settlement of

Coyote Wells in California marks a place where coyotes were the first to find water.

Coyotes naturally do not distinguish between domestic and wild animals, and they will on occasion take poultry, lambs, calves and piglets. This has caused them to be persecuted on a large scale, particularly in the western U.S. Traps, snares, poisons and "coyote getters" were used against them. Coyote getters were scented and baited devices with a cyanide-containing cartridge which exploded in the animal's mouth when it tried to take the bait. Bounties were also paid on some identifiable part of coyotes. In many areas coyotes are still treated as vermin, but there has been no bounty on coyotes in Alberta for over twenty years.

This does not mean that our coyotes have nothing to fear from people. Anyone with a wildlife certificate (the prerequisite to the purchase of a hunting licence for any particular type of game) may hunt coyotes from November 1st to January 31st, that is, for three months preceding their breeding season. Landowners can hunt coyotes on their own land at any time and do not need a certificate. There is no limit on the number of animals that may be killed. Holders of a trapping license may take coyotes in our area from October 1st to January 31st. In spite of all this, the modest coyote population of the Cooking Lake Uplands is not too hard pressed. To my knowledge there is little or no trapping in the area and, in over twenty years, I have only once come across a coyote which had been shot. Our district game warden tells me that only rarely does he hear complaints of poultry-stealing or attacks on young livestock by coyotes, and he believes that even some of these complaints are not justified. He says that most local farmers favour the coyote because the benefit they derive from its rodent hunting outweighs its occasional harmful activities.

I have watched a coyote catching two ground squirrels in quick succession. His technique was to stalk them while keeping as low as possible until he was within about 10 yards of his objective. Then he rushed at the gopher and grabbed it in his mouth. The final attack was so fast that while going for the second gopher the coyote rolled right over but had, nevertheless, managed to seize his prey.

When a coyote has detected a mouse, it first becomes immobile then jumps up and comes down with stiffly-held forelegs on its prey. This technique is also used in the winter, for coyotes can detect mice and voles beneath the snow by sound, but without the aid of vision they are probably less consistently successful.

I have also had a coyote watch me hunting. I was waiting for ducks to come flying through the narrows at Hastings Lake at daybreak one October morning, when I noticed a dog-like animal sitting on the opposite shore. Binoculars showed it to be a coyote which was keeping his eyes on me without making the least movement. There was a fair chance that I might hit a duck which would fall nearer to him than to my side of the narrows, or a bird that was only winged might swim toward him. None of this happened in this instance, but it probably pays coyotes, while keeping out of shotgun range, to keep an eye on hunters. In addition to the possibilities mentioned in this case, they may discover ducks hit but not found by hunters.

I have mentioned carrion as a food item. It may in some areas provide coyotes with nearly as much flesh as they get by hunting. Coyotes have learned, for this can hardly be an inborn trait, to make for places where they can see scavenging birds gathering, usually over a carcass. In some areas the birds concerned will be ravens, in others crows. In the same way, coyotes take note of turkey buzzards circling and spiralling down to feed. As far as coyotes are concerned there is nothing special about a human corpse and many of the unburied dead resulting from the American-Mexican war of the last century were devoured by these carnivores.

How does one tell the sex of a coyote in the wild, a prerequisite for understanding its family and social life? I found the answer in the book by Hope Ryden (see sources), who spent a great deal of time studying the animals in the western U.S. One must watch them urinate. Males do so by raising a leg behind or at right angles to the body. If they squat, they pull themselves forward so that the bent legs extend in a line behind, much as in a stallion's urinating posture. Females sit directly over their haunches, or if they lift a leg they always bring it forward.

Coyote bitches are only in heat for four or five days once a year, some time in February or March. The males are in a sexually active condition for three months during the same part of the year. One might therefore expect that having mated with one female, a coyote dog would wander off in search of another bitch, but their sex life is far more orderly than that of domestic dogs. They enter into a pair bond which generally lasts for several years, sometimes for life. The female, toward the end of her two-month pregnancy, digs herself a den. Often she can do this by enlarging a badger or rodent hole, or she may be able to use a natural cavity in rocky country. On one occasion, a litter of pups was found in the skeleton of a horse which was still covered with skin. Often there are other burrows in the vicinity to which the pups may be shifted if the original den is disturbed. Some burrows have a second entrance.

This was the case in a burrow one of my sons tried to dig out. He was planning to take a pup and raise it as a pet. After digging out the burrow for 4 feet (they are often 6 to 9 feet long) he was worn out and tried to drive the occupants out by smoke. Soon he saw smoke coming out of a second hole further up the bank. In view of this he doubted the smoke in the den would get concentrated enough to force the pups out. At the same time it occured to him that they might be too young to make their way out and he might kill them with too much smoke, so he gave up the operation. It was just as well. While young coyotes make engaging pets, as they grow up their wild nature asserts itself and they create problems in an ordinary household.

Coyotes are fertile, generally giving birth to five or six young, though seventeen and even nineteen in one litter have been recorded. Unlike other animals that give birth in burrows, coyotes generally provide no bedding for the pups. Blind and helpless as they are at birth, they live on their mother's milk for the first three weeks. Then they begin to appear at the entrance of the burrow and to take some of the food both parents regurgitate for them. Two or more weeks later they are weaned. When the young are about three months old the den is abandoned, but the family stays together until the fall, though the young are not fully grown until they are a year old and nearly another year must pass before they are sexually mature.

The events involved in reproduction can, however, be more complicated. Occasionally two litters, clearly of different ages and therefore from two different mothers, are found in one den. It is assumed that an older female has permitted one of her grownup offspring to share the den. Whether in such cases the one attendant dog is mated to both females is not known, but it seems likely.

Hope Ryden observed a coyote bitch which had no young of her own but was in a sense part of the family of a breeding pair. This extra female would groom and play with the pups when their mother was away from the den and discretely retire on her return. She acted, in fact, as a baby sitter and also helped the mother coyote to move the pups, one at a time, to another den. The three animals mentioned were part of what may be called a pack of seven coyotes, consisting of a dominant male and female, the one which had young, and other males and females, all of which at times brought food to the pups. The subordinate animals sometimes appeased the others by lying on their backs in front of one of the "bosses". Such an appeasement posture is also seen in wolves. It inhibits the aggressive urge of the dominant animal and is important in preventing killings within the species of animals which are particularly well armed.

Packs of this size must be unusual, for I have never seen more than two full grown coyotes together in the Cooking Lake Uplands.

Not only are coyotes good parents. It is claimed that, on a higher moral plane, they will help one of their kindred in misfortune. Coyotes have reportedly been seen bringing food to trapped or seriously injured animals of their kind.

In prairie country, a coyote is sometimes seen keeping company with a badger, apparently for the mutual benefit of both parties. While the badger is digging out a ground squirrel or gopher, the coyote may catch one of the rodents as it pops out of some other hole of its burrow. On the other hand, a rodent pursued by the coyote may take refuge in its hole, only to be dug out and eaten by the badger.

Like dogs, coyotes can get mange. One summer I saw a coyote in which the entire rear half of the body, including the tail, was naked. The books mention mange as a possible cause of death in pups, but it seems to me that in our climate an adult like the one I saw might well die from cold in the course of the winter. Like a number of other wild animals, coyotes act as a reservoir of rabies and may at times transmit the disease to domestic animals and potentially to humans if an infected coyote were to bite a person. According to the media, there was a rabies "outbreak" in Alberta in the early 1950's, but to my recollection not one case in humans was reported.

The relationship between domestic dogs, coyotes and wolves is much closer than one would suspect in view of the considerable outward differences between these animals. Coyotes can breed with dogs and the resulting hybrid offspring, coy-dogs, are themselves fertile. Similarly, coy-wolf hybrids have been bred, and these too are fertile. These crosses are obtained with captive animals and must be extremely rare in the wild, if they occur at all. That cross-breeding between, say, dog and coyote can occur is not surprising, but that the hybrid offspring of such a union are fertile and will breed true argues for an extremely close relationship between the parent species.

The Cooper's Hawk At The Nest

The red-tailed hawk is our most common bird of prey. It is large and imposing and is most often seen soaring high over the countryside, its long broad wings hardly making a move. As it goes into a turn, binoculars reveal the rusty red on the upper surface of the tail, from which it gets its name. When the leaves are off the trees, it is easy to spot the large masses of sticks high up in some tall trees which constitute the nests of these hawks. They generally use the same nest year after year.. If one has marked the situation of a nest in winter, one can find it again in the spring, though it may be largely hidden by leaves, and chances are that it will be in use. As one approaches the hawk may drop out of the nest, regain level flight and quietly fly away, or it may circle over the immediate area uttering a plaintive cat-like "pee a". While it is not too difficult to find an occupied red tail nest, it is not so easy to check its contents. Almost always the lower part of the trunk of the nest tree is without branches, and climbing irons will be needed to look into the nest.

I've never come across any hawk nests that were easier to find than those of Swainson's hawks in the prairies of southern Alberta. These close relatives of the red-tailed hawk also breed in trees. However, in that country trees are so few and far between that there is a nest in just about every tree. All you need to do in the breeding season is check one or two trees and you'll find a Swainson's hawk nest.

Things are quite different with that secretive woodland raider, the Cooper's hawk. It is a greyish, crow-sized bird with a long tail and short rounded wings which enable it to slip with amazing agility through the tangle of branches in the woods, to come on some unsuspecting bird and snatch it off its perch. When I first came to Alberta this hawk was believed to be absent in the province or at least to be extremely rare. But now, no doubt because there are more people interested in observing birds, we know that it is actually fairly common in our parklands.

From time to time we would see a Cooper's hawk flying out of the bush into the open near our cottage and continue its flight till it was out of sight. Often smaller birds would fly after the hawk, darting at it in the mock attacks which are known as mobbing. On several occasions we could see that the hawk was carrying prey in its talons. As it was the breeding season, this meant it was carrying food to the female on the nest or to its young. I noticed that the hawk was consistently flying southwest on these prey-carrying missions, and therefore determined to search in a sizeable wood which lies about half a mile away in that direction. Cooper's hawks build rather small nests in trees of average height well inside the woods. These nests are not easily spotted, even in winter, until one gets very close to them. Moreover, finding a nest at that time of the year is of little use since the birds almost always build a new one every spring.

Several times early that summer I wandered about in the wood in question but saw neither the hawks nor their nest. That did not prove they were not there, for a female Cooper's hawk on the nest can spot an intruder from her elevated perch when he is still some way off. She then slips away silently and vanishes among the tree tops, not to return until the coast is clear.

It was not till an evening in early August, when the light inside the wood was already dim, that I disturbed two medium-sized birds that flew off among the trees with a characteristic call. At the time, I thought they might be young long-eared owls, but my next visit showed what they really were. On that occasion I heard the same strange call again and followed it through the thick undergrowth. As I emerged onto a more open area there right in front of me, perched on a slender tree that was arched over like a hoop, was the adult female Cooper's hawk cackling at me in apparent defiance. After a matter of seconds she took off, but high up in a poplar I saw a smaller, browner hawk, obviously one of her young, and glimpsed another similar bird flying between the trees.

It was not until November that I discovered the nest in which these youngsters must have hatched in a sizeable poplar almost beside the bent-over tree.

Next year I naturally kept my eye on this particular wood. Once in late April I saw the pair of hawks circling at some height above it for a few minutes. This was significant, for these birds rarely fly above the tree tops and they are reported to do this circling over the area they have selected for their nest. Nevertheless, the secret breeding habits of these hawks were hidden from me until late July. By that time the young are almost full grown and it seems their mother becomes more demonstrative towards an intruder. At any rate, as I walked along the only trail through the wood on a sunny afternoon, I suddenly heard the cackling call of the female above me and saw her taking short flights from tree to tree as I moved about. After walking short distances in different directions from the spot where I'd first seen the hawk, I found the nest. It was about forty feet up in the fork of a poplar. Two of the young sat bolt upright on branches above the nest and one youngster was still in it. The nest was a mere heap of sticks, flattened at the top and about eighteen inches across. It was a replica of the nest of the year before, which was only about three hundred yards away. I found no remains of prey at the base of the nest tree but on a later visit there was part of the skeleton of a ruffed grouse and a dead vole.

On my second visit I watched the nest for a long time, sitting on the ground with my back against a tree trunk. With binoculars I could make out that the young hawk in the nest was pulling bits of red meat from something he held in his talons. This kept him busy for a full half hour so it must have been a large item, most likely a ruffed grouse. When he'd taken his fill, he climbed a bit above the nest and rested there contentedly. Suddenly the adult female flew onto the nest and immediately left again. Almost certainly she'd brought in more food, but at the angle from which I was watching the bulk of the nest prevented me from seeing this. In any case, the satiated youngster just stayed where he was. The hen, on leaving the nest, did not leave the area right away but perched fairly low on another tree. This gave me a chance to see her gleaming orange-red eyes, her long banded tail, and to admire her elegant slim shape.

Later I found that there were actually four young hawks. Three were perched on branches near the nest tree and one still remained in the nest. By early August they had all moved away some distance, but I could still find them by listening for their whining "peee" hunger calls. I last saw them on the 9th of August. By that time they were flying well and no doubt soon afterwards they dispersed to hunt on their own.

The following year, a journey abroad prevented me from looking for the Cooper's hawks. The year after I was able to find their nest at the very beginning of the season by sheer luck. On May 1st I was walking near the nest in which the pair had reared young two years earlier, when a female flew onto it. We must have seen one another at the same time, for within seconds she was off, cackling as if annoyed. She landed in a tree some 80 yards away, cackled again, moved on to another tree, cackled once more and then flew out of sight. Birds seen through binoculars always look larger than they really are. A female Cooper's hawk is barely 16 inches long (males are smaller) but during the seconds the bird stood on the nest with wings partly raised she looked powerful and imposing.

Disturbance early in the breeding cycle is likely to make birds of prey give up their nests, so I was careful not to visit the nesting area for two weeks.

When I went there again, the nest, forty feet up in the fork of a poplar, had evidently been built up some more but there was no hawk in sight. As I waited some distance from the nest for the female to appear, I heard a "ga-ga-ga" call, fainter and softer than the female's. Walking in the direction of this sound brought me to the edge of a swampy clearing and there on a slim tree, arched over so that its middle was horizontal, sat the male Cooper's hawk. Apparently it didn't notice me among the trees, for it remained undisturbed. Binoculars showed its dark crown, whitish face with fiery red eyes and, from the chest downwards, vivid alternate rusty orange and white stripes. The bird's back and wings were hidden from me, but they are bluish grey. The male Cooper's hawk was altogether more colourful than his mate, with her subdued browns and greys. Her colours were, however, quite atypical. Adult Cooper's hawks of both sexes are normally equally colourful and differ only in size. The female, as in most raptors, is larger. This particular female had retained a brownish plumage through the years I observed her. Perhaps it was a throwback to an ancestral plumage for in a close relative, the European sparrowhawk, females are dull in colour compared to the males.

Next day, as I walked towards the same spot, the male hawk rose up from among the tree trunks and settled in a nearby tree, again in full view. This time I noticed a medium-sized tree broken about three feet from the ground with the long upper portion slanting away from the break to rest on the ground at its extremity. This formed a perch for the hawk more convenient than the branch of a live tree. Below the broken tree I found feathers from a sandpiper and from a robin-sized bird, as well as a piece of mouse skin. Evidently this was the male hawk's plucking log to which he brought his prey for removal of the feathers. After roughly defeathering his prey, he would eat the head and entrails and pass the rest to the female, who spent most of her time on the nest. Of her I had a mere glimpse that day. The nest looked quite deserted, yet as I came closer she dropped out of it, flying down at an angle of 45°, only to vanish among the trees. A week later I found the remains of a red-winged blackbird, an almost complete skeleton of a teal and some fur from a snowshoe hare by the male's plucking log. On May 29th, my younger son, using irons, climbed to the hawk's nest and reported four white eggs the size of hen's eggs. At this time and again later the male hawk had brought a robin to his log. The female was noticeably more reluctant to leave as the process of incubation continued. Whereas earlier on she flew out of the nest when the nest tree was approached, she would now only do so when the tree trunk was tapped with a stick.

Most birds of prey do not hunt smaller birds in the vicinity of their nests, so that the nest area becomes a sanctuary for potential prey. Thus, a red-eyed vireo was in song right through the season close to the Cooper's nest, and an oven bird usually sang nearby as well. On one of my visits, I disturbed the female Cooper's hawk from the plucking log, on which the rear half of a sora rail was lying. The male, who had brought it in, had probably taken what he wanted from it, called the female off the nest, and my appearance had prevented her from finishing her meal.

One evening when I visited the hawks, there was a porcupine fifteen feet up on a dead tree trunk not far from the nest. This might have been expected to alarm the breeding hawk, but she was on the nest, invisible as usual, till I tapped the tree. On June 23rd, a friend climbed the nest in order to band the young hawks, but he found he was much too early for this. The nest held three small balls of white down, youngsters which were only a day old, and an egg which had cracked from the efforts of the last chick to emerge.

From this time on I never found any remains of prey, apart from scattered feathers by the plucking log, and I suppose the female must immediately have taken everything the male brought to the nest for the young and herself. On the undergrowth below the nest the circle of droppings gradually grew wider, but it was not until July 10th that I could see one of the young hawks, still in white down, from the ground. By the 25th they were almost completely feathered and it was evident that only two of them had survived. When I looked for them again on the 2nd of August, both were out of the nest, perched on branches a few feet above. At my approach they flew off, cackling much like their mother. They were, however, still dependent on their parents for food. A few days later I found them near the plucking log, where there were now feathers of a bluebird and of a yellow warbler.

This year as I was patrolling the wood in which the hawks had nested previously, a Cooper's hawk, no doubt the female, flew out of the nest she had used the year before. Again I postponed further visits for two weeks. My disappointment may be imagined when on my next check of the site a red-tailed hawk flopped out of the nest. A pair of these large hawks had usurped the nest and enlarged it a great deal. They reared their young in it in due course. For the rest of the season I was unable to find the Cooper's hawks.

Lynx

The Lynx

The lynx and its close relative the bobcat are among the most powerful predators for their size. Both can kill a full grown deer weighing a hundred pounds, yet these cats weigh only thirty to fifty pounds at the most. It must, however, be pointed out that both species normally live on much smaller prey.

There may be a very few lynx in the Cooking Lake Uplands at all times but my observations suggest that they only appear there as wanderers at approximately ten year intervals. In 1962 my wife and one of my sons saw one on our acreage. It was sitting in the open near the edge of a pasture and at first contemplated them calmly. Then it walked off into the nearby bush and vanished. I was taking a nap at the time and missed this rare sight. By the time they showed me the spot, the animal could not be found. Ten years later I found lynx tracks on the shores of Hastings Lake. During the same winter a friend saw two lynx in a tree trying to catch a squirrel. One or two country neighbours told me about catching glimpses of a bobcat that winter. They had actually seen a lynx, however, for the bobcat is restricted to the very south of Alberta. The Cypress hills are its major stronghold there. About the same time a neighbour's son shot a lynx. Ten years after that, in 1982, though I saw no lynx signs, a friend found the tracks of one.

My explanation of their local appearance at about ten year intervals is as follows. Snowshoe hares are the staple diet of the Canada lynx, though it also takes game birds, particularly spruce grouse, mice and voles. It is well known that snowshoe hares build up to very high numbers every nine to ten years. Peak numbers are followed, generally in the next winter, by a so-called crash, in which the "rabbit" population shrinks to a fraction of its former size and corpses of hares which have died without obvious cause are found. When a rabbit crash has occurred, particularly if grouse and the small rodents are also at a low, lynx in search of prey are forced to wander further afield than normally. The main habitat of the lynx, the boreal forest, begins about a hundred miles north of the Cooking Lake Uplands. Some of the lynx which are facing short commons on their home ranges are quite likely to wander this far south, particularly as they may range over 25 miles even when things are normal. With luck, wandering lynx may come into an area where the "rabbits" are at a different phase of the cycle or where at least small rodents may be fairly common.

The dramatic effects of a rabbit crash on lynx which remained in the boreal forest were recorded by the writer Ernest Thompson Seton (1912), who in 1907 travelled down the Athabasca River to Great Slave Lake and beyond with the zoologist Preble. It was the spring following a snowshoe hare crash. The travellers saw almost no "rabbits" but quite a number of lynx. This is unusual, for the lynx is normally very secretive and restricts its activity to dusk and dawn. The lynx they saw were forced by famine to a continuous search for food during which they were seen in clearings and on river banks. About a dozen were collected as specimens. All were emaciated. Most had empty stomachs, some of the stomachs containing bits of leather or rope which the animals had swallowed in attempts to relieve their hunger. Only two of the stomachs contained food, one a mouse and the other a chipmunk.

Lynx and bobcat are certainly very closely related yet they are distinct species and where their ranges overlap along the U.S. Canadian border they do not interbreed. Both are large, rather long-legged cats with short tails and ear tufts and, in both, the long hairs where the head passes into the neck form a sort of ruff. Bobcats are generally somewhat reddish brown while the lynx tend to be grey, but the only reliable difference, not at all easy to see in the wild, is in their tails. The tip of the tail is all black in the lynx. Bobcats have a tail tip which is black only on top but white

Bobcat

underneath. Usually they also have shorter ear tufts. The bobcat is distinctly more southern in distribution than the lynx. It ranges from southern California, including all of the maritimes, to Florida and well into Mexico. The lynx is almost entirely Canadian in distribution though along the Rockies its range includes part of the western states. Most modern zoologists consider that our lynx is of the same species as the animal found in Spain, the Balkans and from eastern Europe and Scandinavia right across Siberia. Lynx and bobcats have spots but in the Spanish lynx and that of the Balkans, often called the pardel lynx, the spots are very distinct.

Old World lynx have the same propensity to kill deer as the Canada lynx and the bobcat. One of the French names of the lynx is *loup cervier*, literally "deer killing wolf" and this is also the meaning of its Italian name.

To understand how a lynx or bobcat can overcome a much larger animal one must recall the amazing transformation even a domestic cat undergoes when cornered or attacked. Docile pussy becomes a fearsome beast raging with tooth and claw. There are few eyewitness accounts of lynx attacks on deer but the few there are, and the greater number of reconstructions based on finding the dead deer or its remains and tracks in the snow, give much the same picture. Attacks on full grown deer almost always take place in winter when lynx are most hard pressed for food. In snow, particularly crusted snow, the lynx is at an advantage to the deer for

59

its wide paws in their winter fur act like snowshoes while the deer's hoofs break through the crust.

The predator may lie in wait on a limb above a game trail and leap, claws out, onto the shoulder of a deer passing below. It attacks immediately with teeth and claws, tearing at the large blood vessels of the neck. Once one of these is torn the deer very soon collapses as a result of hemorrhage and shock. Almost always the grip of the predator's claw is so firm that it is not dislodged by the deer's frantic leaping about nor is it foiled when the deer flings itself down in an attempt to push the cat off. The cat holds on. Should it perchance be thrown, it leaps straight back onto the prey.

More often the lynx stalks a deer which is feeding or lying down in the snow. The predator approaches, taking advantage of every bit of cover, in a series of crouching crawls until it is within about five yards. Then it makes a tremendous leap, again aiming at the shoulder of the deer. If it misses the cat gives up. Although it is capable of short fast runs, these cannot match the speed of a full grown deer. A lynx's run, really a series of bounds, actually looks rather clumsy as the long hind legs raise the animal's hips a bit above the shoulders. Its "long jumps" however are impressive and very swift. A Spanish Lynx has been seen to claw a red-legged partridge, out of the air.

In the absence of snow, deer are not always at the mercy of these predators. There have been several reports of does routing a bobcat with their formidable front hooves as it was sneaking up on one of their fawns.

An observation which shows the amazing urge of a mother bobcat to take prey in the face of danger is related by Stanley Young in his book "The Bobcat of North America". His witness saw a female bobcat, obviously alarmed, carry her kittens one at a time up into a tree as a half wild boar followed by a sow and her litter entered the forest glade where all this took place. The boar, on catching sight of the cat, rushed at the tree, foaming at the mouth and gnashing its five-inch tusks. Unable to get at the cat it gradually calmed down. The mother bobcat, which shortly before had sought refuge from the boar, now began stealthily to inch down the tree. Finally she sprang at one of the piglets and carried it off, being immediately pursued by the boar.

The bobcat rushed toward a ledge of rock, throwing the piglet aside when its weight hindered her upward scramble, and got away from the boar. Not long afterwards the pigs left the area. The bobcat then returned to the tree and brought the two youngsters down to the ground, carrying them, as before, in her mouth.

After killing a large animal, lynx and bobcats eat only a small part and cover the rest of the carcass with snow or earth. It is not certain whether they come back and make full use of what is left. Perhaps they do if it is still fresh and if they don't come across other food, but these cats, in contrast to wolves and coyotes, practically never eat carrion.

It is more common for the lynx to attack a young deer than an adult. It will also take young caribou and even, as reported from Scandinavia, moose. In Europe the lynx will attack young of the small roe deer and, in mountainous areas, those of chamois. Attacks on domestic animals are fortunately rare but include sheep, goats, piglets and, very rarely, foals.

Lynx attacks on big game are most likely to succeed if the prey is not quite up to the mark due to sickness, old age or injuries. Perfectly healthy animals are also taken on occasion, but various studies in Europe have shown that the overall effect of lynx predation is a weeding out of the unfit among the prey animals. A Swedish survey of lynx-killed roe deer showed that 60 percent were diseased. In the Polish Carpathian mountains the post war increase in the lynx population was accompanied by an increase in the number of roe deer. This was attributed to a weeding out of the unfit roes. Red deer and wild boar also became more numerous. Lynx cannot overcome adults of these two species but they prey on the young and among these they apparently most often take the unfit. In the Bielowicza reserve in Eastern Poland, where the last European bison are preserved, there are about a hundred lynx. They kill two to three hundred roe deer every year and these are judged to be mainly the sick or weak, one tenth of the local roe deer population. The lynx must therefore be judged to have an overall beneficial effect on big game populations.

Canada lynx occasionally feed on foxes. This happens in winter when a fox with its narrow paws sinks in soft snow and soon gets exhausted as it leaps along to get away from lynx. The latter with its wide heavily furred paws (those of the hindfoot may measure five inches across) can move far more easily over the snow and before long catches up with the fox. The brief fight which follows is always won by the lynx.

The fact that the Canada lynx is so largely dependent on the snowshoe hare results in fluctuations of the lynx population which follow those of the hares. The records of furs bought for resale by the Hudson's Bay Company over a period of 1 1/2 centuries show peak numbers for the lynx at about ten year intervals. The peaks coincide, as we know from recent observations, with peaks in the snowshoe hare numbers. Not only does the lynx population decline from starvation when rabbits are very scarce, but also because they raise fewer and often no young during these periods. On the other hand, when the prey is abundant just about every female lynx gives birth to two to four young. The same phenomenon is seen in snowy owls and rough-legged hawks both of which prey very largely on lemmings. Lemmings build up to a population high every four years. In lemming years both these predators lay more eggs than usual while they may lay none at all in years of lemming scarcity.

One may speculate on how prey abundance stimulates fertility. It is well know that the function of the sex glands is stimulated by certain hormones of the pituitary gland. This gland is in turn controlled by hormones from the hypothalamus, a part of the brain. Quite possibly the frequent sight of and contact with prey by the predators stimulates their hypothalamus with a consequent activation of their sex glands and reproductive systems.

The bobcat, unlike the lynx, is not dependent on a particular animal species for its food and none of its prey animals fluctuate in numbers like the snow shoe hare. It is therefore not surprising that the yearly catch of bobcats remains fairly steady over long periods and does not show the cyclical pattern of the lynx harvest.

Lynx lead solitary lives outside the mating season which occurs in early spring. The sexes are probably brought together by scent and the loud nocturnal caterwauling of the males at this time. Males fight a great deal when several are courting the same female until one of them becomes the chosen suitor. Towards the end of a pregnancy of about nine weeks the female selects a den which is protected from wind and rain. It may be in a hollow log or under the roots of a fallen tree. The male is expelled from the den when the kittens are born but he generally remains in the vicinity and brings food to the entrance until the young can follow their mother about.

Every mature lynx normally maintains a territory of his own and marks it by leaving his droppings in prominent places in the open. Males also use particular trees to sharpen their claws and such trees perhaps also function as territorial marks.

In Europe (though apparently not in North America) the expression lynx-eyed, meaning sharp-eyed is frequently heard. The idea that lynx have particularly good vision probably stems from the fact that for an animal of the cat tribe they have rather large eyes.

Some experiments conducted in Germany on lynx reared in captivity show that their vision is in fact better than ours. These lynx could spot a large hawk in flight two miles away. When a stuffed hare was moved on snow by pulling a string attached to it they noticed it at a distance of 325 yards. Their hearing also proved to be superhuman. They reacted to a police whistle at a distance of 2.8 miles, while people could only hear it up to 1.5 miles and dogs up to 1.8 miles away. The ear tufts apparently function in connection with hearing for when they are cut off the lynx's ability to hear and locate sounds is reduced. The tufts probably act as non-directional detectors of sound waves, the source of which is then located by rotating the ear.

Common loon on nest
Photo by A.E. Weedman

Loons

Common loons have nested in Elk Island National Park at the northern extremity of the Uplands but in the rest of the area they only appear as occasional spring migrants. When the lakes are still frozen in April I am likely to see single loons, or occasionally two together, on one or other of the smaller waters which are already ice free. A day or two later these conspicuous visitors have left. They are apparently on their way to the forested areas further north where countless lakes are the breeding places of most of the species. One year a common loon remained on Hastings Lake until the end of July, but that was quite exceptional. Non-breeding birds like this loon often spend the summer well to the south of the breeding range of their kind. Loons are not sexually mature until they are at least two years old, so this bird may well have been too young to breed.

The call of the loon is justly famous. For so massive a bird it is surprisingly musical and can be heard for several miles. I shall always remember the first time I heard it. The late Professor Rowan and I were wandering through a northern Alberta muskeg on a beautiful sunny day in the middle of May. Suddenly we heard a loud, long, continued, pleasant "song". Its source was hidden from us but it was evidently uttered by a bird on the wing. "A loon" said Rowan, in response to the mystified expression on my face. I knew he must be right but the sound reminded me very much of the European curlew whose bubbling call is now often heard on TV in scenes of the English countryside. What surprised me was that a bird as massive as a loon could yodel so exuberantly and melodiously while flying. Common loons actually have three distinct calls. The most impressive is the yodel. Sounds somewhat like "aa ooo ooo ooee wee oo wee" are repeated several times, each phrase rising in pitch at the beginning and then rising and falling. Loons yodelling while they are on the water usually lower the head and neck and this posture, as well as the call, constitute a mild threat toward other loons. Both are used in the defence of that segment of a lake which a pair regards as its own territory. However loons also yodel on the wing while flying around in wide circles, particularly late in the evening and at night. Probably these flights are mostly made over their territory and also indicate its ownership to others.

The tremolo call is simpler. It consists of a repetition of a note like "wee wee wee wee" on the same pitch and gives the impression of mindless laughter.

Then there is the wailing or howling which is reminiscent of the howling of wolves. Rising in pitch as it proceeds, it falls again in the middle and can represented by a "haa ooo ooo ah." There are also softer "conversational" calls which birds fairly close together seem to use to keep in touch. All these calls are only heard while the birds are on their breeding lakes and sometimes while they are on their spring migration. In winter they are silent except for occasional harsh "kak" calls uttered in flight.

Not all of its relatives call as melodiously as the common loon. The smallest of the loon tribe (there are four species, all natives of the northern hemisphere), is the red-throated loon. It has a long-drawn wail which sounds like a cat's meeaoo and a growling call, justly described as hideous, which sounds like "warr warrooa" repeated many times. Most red-throated loons breed in the arctic but ten years ago I found a number of these birds in the Caribou Mountains of Northern Alberta. Rather than a mountain range, this area is a plateau with many large and small lakes. Common loons live on the large lakes but the red-throats nest only on smaller waters where the bigger loons never come. The pools where red-throats nest have no fish so they fly to the nearest large lake for their fishing. A red-throated loon carrying a fish cross-wise in its beak, flying away from a big lake and over the forest beyond to its hidden breeding place, is a frequent sight in the Caribou Mountains. In Scotland red-throated loons nesting on small moorland lochans generally fly to the sea to fish.

Three red-throated loons gave me a very striking demonstration of territorial behaviour. For several days we had seen a pair on a small forest lake. One evening a third loon flew in, landed on this stretch of water and swam toward the pair. For a few seconds the intruder and the other two birds confronted one another with necks extended forward and heads bent down. Suddenly one of the paired birds rushed at the intruder which immediately dived. After that whenever and wherever it surfaced one or both of the paired birds attacked it. We watched this unremitting pursuit of the stranger for nearly an hour and left while it was still going on. Next morning only the pair was to be seen.

Loons and other diving birds must be able to see clearly under water as well as in the air and some special adaptations of their eyes have evolved in response to this need. To focus on an object under water the eye must accomodate more, i.e. increase its lens power more than is needed to focus on an object at the same distance in the air. Diving birds therefore have a greater range of accommodation than we do. There are no figures on this for loons but cormorants have a maximal power of accommodation four times that of young adult humans. Loons have a very soft lens which can be made globular, when it is most powerful, by highly developed small muscles within the eyeball. In addition there is the third eyelid, or nictitating membrane, which slides outward over the eye from its inner angle when the loon is under water. This membrane has a transparent window which acts as an auxiliary lens.

Diving birds also show adaptations which reduce their consumption of oxygen while diving and others which increase the oxygen store in their bodies. In the first category are an automatic slowing of the heart beat and a reduced blood flow to tissues which, like the intestine, are for the time-being unimportant. The blood volume of diving birds is high and so is the number of their red blood cells. The amount of oxygen-carrying pigment hemoglobin in these cells is also unusually high. The diver therefore starts its dive with a greater reservoir of oxygen in its blood than is found in a land bird of similar size.

Nevertheless, loons do not normally stay under water for a phenomenally long time. Up to a minute and half is average for the common loon. Loons that were being pursued are reported to have stayed under water for 10 to 15 minutes. I take it that "pursued" means that the loon was shot at and missed every time it surfaced and so was forced to stay under water to the limit of its endurance.

Information about the depths to which loons may dive is provided by birds which get entangled in fish nets that have been set at known depths. Dives of up to 200 feet have been recorded on the Great Lakes but usually loons dive only twelve to fifteen feet below the surface.

While fish of many species form the bulk of the loons' diet they also eat frogs and leeches, water snails, shrimps, water beetles, nymphs of dragonflies and other aquatic small-fry. Water plants have also been found in their stomachs, in amounts greater than would have been swallowed accidently with other food. The late J.A. Munro found common loons breeding on fishless shallow lakes in interior British Columbia. He proved by examination of the stomach contents of birds he collected that they had lived entirely on invertebrate animals, such as those indicated above, and water plants. Loons nesting on small muskeg lakes in northern Alberta which have no drainage and presumably no fish must use the same sort of food.

However even these loons have fish available to them in winter. Almost all common loons spend that period at sea in coastal waters from Alaska to southern California and from Newfoundland to the Gulf of Mexico. The migration of western loons must involve a crossing of the Rockies but this is simple compared to the travels of the yellow billed loon. Yellow bills which nest in our Arctic travel westward in the fall along the coast from points as far east as Melville Peninsula, past Point Barrow into the Bering Sea, then southward and finally southeastward to their wintering area off the coast of southern Alaska and British Columbia. At the Bering sea they are joined by others of their kind from eastern Siberia which have travelled eastward along the coast of that country.

Loons usually arrive on their breeding grounds in pairs and banding has shown that the same two birds often return to the same lake year after year. It is therefore likely that they are joined for life. This being so, no elaborate displays are required to initiate mating. The female, nodding her head, swims along the shore looking for an easy place to climb out of the water. Having found one she shuffles along for a few feet and lies down. The male follows her and copulation follows without further prelude, always, as described, on land.

More elaborate displays are used in connection with the claiming and defence of territories. There is the splash dive in which the

Common loon nest

bird throws up water with its feet at the moment of disappearing below the water. This is in contrast to normal dives in which the bird glides under water without any disturbance. More strongly excited loons rush across the water for considerable distances beating it with outstretched wings. Such a rush may culminate in the most impressive display of all, the "fencing posture". In this the loon stands almost vertically erect supported by the frantic paddling of its feet while its head is drawn back as if in readiness for a forward lunge with the beak. Occasionally the bird may even jump clean out of the water.

Most loon nests are within a few feet of a lakeshore, very often on islands. There the eggs are less likely to be discovered by four-footed predators like mink and muskrats than in mainland nests. When artificial islets of matted sedges were set up as nest sites for other waterfowl in Minnesota, it was found that loons often made use of them for that purpose. Typical nests are very simple. A depression is scraped out and surrounded by a ridge of matted vegetation much of which is plucked and deposited by the loon while incubating. Some nests are in old muskrat houses where the loons merely scratch out a depression. In other cases the eggs are laid on a substantial heap of cattails or sedges which is built up in shallow water.

Only two very large eggs are laid and both members of the pair take turns at incubating them for a total of four weeks. The incubating bird always faces the water and is ready to slither into it and dive on disturbance. When the nest is in the shallows among emergent vegetation, an escape route to deeper water where the loons can dive is usually found. The birds may clear it themselves or use a pre-existing muskrat or beaver channel.

During incubation some eggs are lost to crows as well as to mammalian predators. In such cases the loons often lay a second clutch. The young are covered with black down and are brooded on the nest by one or the other parent for the first two or three days of their lives. When they enter the water they are often carried on the back of a parent bird, under its wing, and so kept warm. This habit persists until the young are a little over two weeks old. The young beg for food with piping calls much like those of barnyard chicks and their parents offer them tid-bits at the tip of their beaks. Loon chicks can swim on the day they are hatched but it takes them about a week to acquire the ability to dive. When a pair of loons with young is disturbed the adults generally swim or dive out into the lake and the young hide in the shoreline vegetation.

Loon families start with two young but after some weeks most pairs are left with only one youngster. Recent studies have provided an explanation for this. During their first days of life the young go through a phase of aggression in which the two siblings fight one another in a pose similar to the adults' fencing posture. Sooner or later one of them submits and thereafter the dominant bird secures virtually all the food offered by the parents and the loser eventually starves to death. Curiously enough, ten-day-old young reared in captivity and brought into contact with others have lost the urge to fight and simply ignore one another.

From late summer into the fall common loons gather in assemblies of up to several hundred on certain lakes which apparently act as staging areas in connection with the fall migration.

Though loons are not game birds, and in spite of the fact that most of them nest in remote areas, certain human activities bring about their deaths in great numbers. Throughout their range these birds are liable to be caught and drowned in fish nets. A Canadian Wildlife Service report mentions 5,662 loons caught in one month's annual commercial fishing in Great Slave Lake in the course of 1960 and 1961.

In recent years oil spills at sea, where nearly all the loons winter, have added to the death toll. An example is provided by a spill in the Shetland Islands in 1978 after which 146 dead common loons were found.

Another danger is acid rain. Compounds of sulfur and nitrogen from nonferrous smelters react with atmospheric moisture to form sulphuric and nitric acid which are eventually deposited on land and water. This rain kills fish and must make the lakes concerned useless for loons even if it does not poison them directly.

Finally, loons cannot stand more than a slight degree of disturbance from lakeshore cottages and particularly from motor boats. Lakes subject to such disturbance lose their loon population. Unfortunately this process goes on near all settled areas. In a part of New York State where there were 85 lakes with breeding loons in 1960 there were only 55 loon-occupied waters in 1979, a change mainly due to increased human use of the local waters.

The Tree-Nesting Solitary Sandpiper

Many different kinds of plovers and sandpipers pass through our district every spring and fall. They rest in flocks on the shores of our lakes and larger sloughs and in spring are bound for more northern breeding grounds, in many cases the Arctic itself.

Only the killdeer plover, the spotted sandpiper and the lesser yellowlegs stay to nest in our own district. Not too far away, in the forest muskegs to the north and west of the parkland, three further kinds of shore birds are found nesting — the greater yellowlegs, the dowitcher and that most unusual bird, the solitary sandpiper. While all other shore birds congregate in flocks while on their travels, this dainty bird, dark grey above and white below and up to 9" long, prefers to travel alone — hence its name. Even if two or three of them happen to be feeding close together along some shore, each forms a unit of its own and may fly off leaving the others, who will carry on quite undisturbed. Strangely enough, the solitary sandpiper is the only species which regularly, spring and fall, visits the small waters on our acreage. Surrounded by trees as they are, these waters may remind the birds of their forest homes.

Shore birds nest on the ground where, although their eggs blend amazingly well with their surroundings, they must be exposed to many predators. Not so the solitary sandpiper — it breeds in old nests of rusty blackbirds, cedar waxwings, robins, or even on the bulky structures of gray jays. The male sandpiper finds such nests by hopping about from branch to branch on spruce or even deciduous trees near the pool beside which it has decided to nest. The pair then build up the old nest remnant into a proper nest for their four eggs. Undoubtedly such tree nests are safer from most enemies than are ground nests. But how could such a nesting habit, unique for a North American shore bird, have evolved?

The most likely answer is obtained by considering the habits of its Old World relatives, the wood and the rather ineptly named green sandpiper, a bird which is in fact grey and white just like the solitary. The northern element of the American bird fauna immigrated into the New World during the Ice Ages when a land bridge existed across what is now the Bering Sea linking Asia and North America. Across this bridge in all probability came a bird which was ancestral to all three kinds of sandpipers just named and which in North America evolved into the solitary sandpiper.

In the wood sandpiper the tree nesting habit is only weakly

developed. This bird nests in marshy forests from Holland and northern Germany northwards into Scandinavia and Russia and across most of Siberia. Normally it nests on the ground, but where its chosen sites are inundated by spring floods it uses the old nests of thrushes or shrikes. Its response to flooding is no doubt the origin of the tree nesting habit of this group of sandpipers. In some of its haunts, the wood sandpiper has already become an endangered species, for there are only thirty to forty pairs left in Germany.

The slightly larger green sandpiper has a similar breeding range over much of the northern Old World, but it always nests in trees, using such diverse old nests as those of thrushes, crows, jays and even wood pigeons.

All these sandpipers look rather similar — though the two Old World ones have a white rump which shows up clearly when they fly, while our solitary has a rump of alternate black and white stripes. Their calls too, while distinguishable, are very similar.

The solitary sandpipers on our acreage favour our artificial dugout pond and here they feed in the mud exposed by the falling water level in summer. One year when heavy rains kept the water level so high that no mud was exposed, only one solitary showed up in late June and none were seen during their usual late summer passage season of July and August.

There is a wooden raft on our dugout. Ducks as well as sandpipers often rest on it, for to them it is a little island where they feel safer than on the shore. The sandpipers often sleep on the raft, their heads tucked under a wing. The cover of the shoreline trees then permits an observer to come very close to them undetected, so that every detail of their elegant plumage can be seen.

The solitary sandpiper is easily recognized as different from the few shore birds which spend the summer in this district. Its proportions are elegant like those of the lesser yellowlegs but the latter's yellow legs distinguish it at once from the dark-legged solitary. Yellowlegs are also larger and taller and have a white rump. The spotted sandpiper, about the size of the solitary but more dumpy in outline, is marked during the breeding season by black spots all over its lower surface. It often bobs its whole body up and down, while the solitary only nods its head. While the solitary flies in a dashing manner, the spotted sandpiper flies off with shallow quivering wing beats and shows a white mark on the upper surface of the wing which is absent in the solitary.

When solitary sandpipers leave us in late summer, they move southwards to winter in Baja California in the west, and from

Georgia to Florida and the Gulf of Mexico in the east. Some of them even winter as far away as Argentina.

PART III

Wildlife Through The Year

Wildlife
Through
The Year

To give a picture of Aspen Parkland wildlife as a whole, it seems best to follow its changing aspects through the four seasons.

Of these spring is by far the most exciting to the naturalist. With the emergence of new greenery comes a rebirth of the land — a bear, fresh out of hibernation, lumbers through the undergrowth. The gentle breeze is filled with hundreds of busy sounds — bees buzzing, squirrels chattering and frogs croaking. Migrant birds arrive; some stay, others move on to the far north. Among these birds there are songs, courtship antics, and squabbles to be heard and seen. Early in June this surge of avian passion reaches a climax. For some it is already over, as signalled by the appearance of the first baby ducklings.

By late summer most of our song birds are moulting. The countryside is as beautiful as ever, yet now there is a fullness and ripeness in the air not known by spring. The area is now much more silent and seems almost devoid of birds. Yet on muddy lake shores interesting waders have already arrived from the tundra and subarctic in the course of their southward migration.

In autumn the passage of these birds as well as the later fall migration of cranes and geese continues well into October. The countryside is again beautiful in its robe of orange and yellow, and the crispness in the air warns the squirrel to gather its winter store.

Winter is the quietest season for the naturalist. The landscape may at times be glittering in the sun with snow and ice but it is very nearly empty. Our large mammals are indeed still about but they are few and far between. Most birds have left and though a few interesting species stay through our northern winters, they too are thinly scattered. As one of my friends puts it, "You can go for a long walk and see more birds back home at the feeder than you do in your tramping about." Some patience and above all the certainty of spring to come are needed to make these months acceptable.

This part of the world is marked by the very intensity and distinctness of each season. From the bustle of spring to the stillness of winter there is much to see and experience. Each season holds its own surprises for anyone prepared to discover them.

Spring

For the naturalist spring begins with the arrival of the earliest migrant birds when the countryside is still in the grip of winter.

Horned Lark Horned larks, often seen in the middle of February, are the first to appear, generally in small flocks along roadsides where they can find weed seeds above the snow. I have heard their cheery tinkling song as early as the first of March. Wooded country does not attract them so they avoid the Hastings Lake area but they breed in the fields and pastures of open country.

From late March to early May one can hear a sound in our woods much like that of a distant motorcycle starting up. It is pro-

Ruffed Grouse duced by ruffed grouse cocks drumming. At close quarters the drumming sounds like "wup wup whurrr". Snapping wing beats back and forth cause the "wup" sounds; they lead up to very fast beats which so compress the air that finally a sound like the muffled roll of a drum is produced. While drumming a cock stands on a fallen log. He may have two or more drumming logs in his territory. As a cock uses these for the entire drumming season, favorite logs are marked by an impressive pile of droppings.

The drumming has two purposes. One is to lay claim to the surrounding territory, warning off other cocks and suggesting they go elsewhere. On the other hand it attracts hens. On seeing another grouse the cock leaves his log to display to it on the ground.

The cock's ground display is much the same whether the visitor is a hen or another cock. He struts toward the other bird with slow dignified steps, with raised head, expanded ruff on the lower neck and the tail with its beautiful bands spread wide. Repeatedly the cock bows and shakes his head uttering a hiss with each bow and finally he comes to a veritable climax of bowing and head shaking. He then makes a run at the other bird with wings dragging on the ground. If the visitor is a hen she will either withdraw a little at the cock's approach or crouch down with raised tail, in which case a mating will take place. The two partners only stay together a few hours for the hens soon leave the mating area in search of nest sites. The male stays in his drumming area and mates with any other receptive hen attracted by his drumming.

If a cock wanders into the drumming area of another male, both birds assume the ground display posture and threaten one another with raised ruffs and outspread tail. The intruder rarely feels as confident as the owner of the territory and is soon intimidated. In the rare instances where he maintains his ground, the issue is decided by a short bloodless fight.

Drumming can be heard for half a mile, so a long search may be needed to locate the particular cock that produced it. When found, the bird, having heard the observer's approach, is most likely seen running away from his log in a crouching posture.

The drumming display of ruffed grouse has often been photographed and filmed from a blind, but one spring I found a cock close to our acreage which could be observed without any such device. His drumming log was only thinly screened from a trail. On the other side of the trail was a dense growth of young poplars so that standing among them, unnoticed by the bird, one could watch his performance. He would stand on his log, the crest on his head slightly raised, ruff spread out, wings hanging at his side and tail expanded. Beating his wings he would produce the preliminary "wup" sounds and then, with wing movements too fast for the eye to follow, make the climactic rumbling sound.

Drumming is most often heard in the early morning and again in the evening, but on the day I found this particular male he drummed every five minutes, even around noon.

Close to the spot where this cock had his headquarters there is a secondary road. One day early in May I found this bird, obviously traffic killed, lying dead on the road. However, it had left descendants. Later that summer two hens, each with a brood of young, showed up in the area where it used to display.

Young grouse can fly when only ten days old and still quite small. When a brood is disturbed the hen flies off to a nearby perch or out of sight. The chicks, mere little balls of down and feathers, erupt into the air on whirring wings and zoom off in different directions. They hide wherever they land and are almost impossible to find. When the disturbance is over the hen with soft calls gathers them together again. By late September the family group has broken up. Ruffed grouse are then encountered singly or at most with one or two companions. The young, now twelve weeks old and full grown, can no longer be told from older birds.

Ruffed grouse are now virtually the only upland game bird of our district. This was not always the case. Neighbouring farmers

Sharp-tailed Grouse

tell me that until forty years ago there were as many sharp-tailed as ruffed grouse on their land. Thirty years ago I myself watched sharp-tailed grouse on a dancing ground near the land where the Alberta Game Farm was later set up. About twenty years ago

Gray partridge nest

my oldest son found a brood of chicks on our acreage. During the following few years we occasionally flushed one of these birds and heard its characteristic cackling call as it flew away. But we have never come across any more of them in the Hastings Lake district in recent years. Their disappearance here is only a local phenomenon for they are still to be found in the more open country to the east around Beaverhill Lake, and populations are found throughout the open parkland.

The gray partridge (locally known as the Hungarian partridge after the country from which some of these European birds were

Hungarian Partridge

introduced) is essentially a bird of cultivated fields and hedges. It is unlikely that it ever became common in our fairly wooded districts. Nevertheless a farmer once showed me one of these birds on the nest in a field ten miles from our acreage. But that was many years ago and I have not seen a partridge in our area since then. Their local disappearance is part of an unexplained decline which has taken place over a much wider area in the last twenty years or so.

Pheasants have also shown a similar widespread decrease during the same period. The few pheasants that could occasionally be

Pheasant

seen in the Hastings Lake country, particularly along its margins, have disappeared. Most likely both the pheasant and the partridge, which are less hardy than our native game birds, were adversely affected by a number of severe winters.

By late March mallard and pintail ducks begin to arrive. They are first seen on the shallow pools formed by the spring run-off at a time when the permanent waters are still frozen.

At first mallards and other dabbling ducks are assembled in flocks but these soon break up into pairs which scatter widely as

Mallard Duck

more and more waters become ice-free. These pairs were actually formed earlier while the ducks were still in their winter quarters. There they indulge in group displays in the course of which the females choose their partners. It is a bond which lasts well into the coming breeding season.

Once mallards have spread over the countryside in early spring they indulge in much frantic flying about in small groups. The frenzied quacking of a hen mallard is heard. Seconds later she comes winging over the tree tops closely followed, one behind the other, by two drakes. Trying to shake off her pursuers the duck climbs or swoops down. She may fly a tight curve to one side or the other but the drakes follow her every move. The leading drake repeatedly tries to grasp her tail. She has good reason for her excited calls. The second drake on the other hand tries to bite the tail or feet of the male he is following.

Because of the many scattered woodlots in my area the birds are usually lost beyond the trees before the end of the flight. But if they remain in view long enough, the drake immediately behind the duck is seen to turn away and fly on in a more leisurely manner. The duck, followed by the second drake who is actually her mate, then continues on her way.

When these three-bird flights are studied, as has been done on this continent as well as in Europe, their meaning becomes clear. They arise when a duck followed by her mate comes too close to a pair already in possession of a site. The "landowner" drake then immediately chases her and he is followed by the mate of the intruding hen. After a variable distance of pursuit the "landowner" drake sheers off and returns to his female while the prospecting pair eventually alight elsewhere.

Somewhat later in the season, from mid-May onward, there are pursuit flights in which a greater number of males chase a single female in flight. They are mostly drakes whose pair bonds ended when their hens were well into incubation. In these sexual pursuits the mallard hen has every reason for her fearful behaviour for if she is forced down by a male flying above her, she may be raped by one of the pursuing drakes.

There is still another form of pursuit flight involving birds not yet paired. These flights seem to be part of a pair forming process taking place in spring and not, as is usually the case, in winter. This is suggested by instances where, after a pursuit on the wing, mating is clearly not resisted by the hen.

Canada geese come in about the same time as the ducks and are often content to manage without water. They will settle on partly snow-covered fields or on lakes that are still **Canada Goose** covered with ice. To hear the loud challenging call of these geese from the far side of a belt of naked poplars and then to see a pair of the large birds appearing over the tree tops, conveys the feeling that life has indeed returned to the still wintery land. This is a common experience in early spring around Hastings Lake where later in the season a number of pairs of Canada geese settle down to nest on the islands in the lake.

Canada geese are followed by whistling swans in early April. These are on their way to breeding areas along the coast of mainland Arctic Canada and Alaska. By late **Whistling** April their spring passage through our dis-**Swan** trict is concluded. Generally they are seen in flight after their musical "oo boo oo" calls are heard from afar. Whistling swans weigh up to twenty pounds and make a magnificent sight with their pure white plumage against a blue spring sky. Sometimes they fly quite low. Then the swishing sound of their powerful wing beats is added to their melodious calls. As well, their dark beaks and legs become evident. Many, but not all, have a small yellow spot at the root of the beak. This is only visible at close range and will distinguish them from the larger trumpeter swan. The only trumpeters that might migrate through our district are the 150 or so which nest near Grande Prairie. These probably use a migration route well to the west of our area. Should a trumpeter appear in our district it could only be distinguished by its larger size, or by its louder, more resonant trumpet-like calls.

Like geese, swans mate for life. In the fall, when whistling swans are seen in greater numbers than in the spring, individual families can be made out among the flocks. An adult will be leading two to four pale grey young swimming in line behind it, with another adult bringing up the rear.

In 1955 I was able to spend July and August at the mouth of the Anderson River in the western Arctic. The delta of this river, with its low-lying islands, is a breeding ground of snow and white-

Trumpeter swan
Photo by Ron Mackay

fronted geese as well as whistling swans. It has since become a waterfowl sanctuary. I had arranged for the pilot and owner of the Aklavik flying service to fly me to the Anderson, where the cabin of a former reindeer herder was to be my headquarters on June 1st. However when I reached Aklavik I was told my pilot was still in Edmonton. So it was not until the first day of July that he delivered me, along with the eskimo dog I had picked up at Tuktuoyaktuk and my two months food supply, on the river below the cabin. The dog, which I called Kinga (eskimo for "the nose"), proved so good a companion that I never felt lonely during the weeks which followed. He was very clearly aware of his dependence on me for food and always kept close at hand. When we went further afield I made him carry my small tent and some of our food in the pack sack I had made for him.

The swans were already breeding and I found my first nest on the day after my arrival. It was a mass of brown peat, with a few feathers among the three eggs, built on a ridge which carried some low willows and separated two ponds. A female was incubating when I spotted the nest. I was still about 60 yards away when she left, took to the water and flew away. As I was inspecting the nest she flew over low as if to check what was going on. Later in the month I found another nest at the very edge of the tidal flats on the estuary of the river. The first downy young were seen with their parents on July 28. At the beginning of August the swans were flightless, for like all waterfowl and loons they shed all their flight feathers after breeding. On the eighth of August I noticed many moulted swan feathers lying about on the islands in the river below the cabin and when I approached some swans on a small lake, instead of flying away, they walked ashore and marched off to the nearest larger lake. By the 22nd of August most of them were once again able to fly.

In this and other arctic breeding areas these swans have virtually no enemies. Golden eagles were more common about the Ander-

Golden Eagle son than I have seen them anywhere else but they always hunted over escarpments on the coastal slopes. Here there were burrows of Arctic ground squirrels and these were evidently their favorite prey. It seems that neither golden or bald eagles have been known to kill birds larger than geese. Swans are too large for them and they leave them unmolested. Arctic foxes are eager egg thieves but swans can certainly keep these marauders away though they might not be successful against that determined predator the wolverine. Wolves are at their dens during the bird breeding seasons and are far from the marshy areas where the swans nest. In all likelihood eskimos, who may occasionally shoot swans or take their eggs, are the most serious enemies of swans on their breeding grounds.

Another, though far less obtrusive, early spring arrival is the mountain bluebird. The intensely blue males are generally first

Mountain Bluebird noticed as a flash of colour when they move. Their call, a pleasant soft throaty whistle, also draws attention to their whereabouts. These small thrushes, which originally nested only in natural tree holes, are equally content to breed in nest boxes. A number of such boxes have been nailed to power poles around my home area and, though most of them are used by tree swallows, several are occupied every year by the more glamorous bluebirds. A pair breeds on our acreage in one or another nest box nearly every year. Immediately after the young of the first brood have left the female lays another clutch of eggs and the birds generally succeed in raising a second brood before the end of July. When these birds leave us in the fall they may go no further than Montana but, as they occur in winter in all the western states and in Mexico, many must travel to far more distant winter quarters.

In nearly all birds both sexes utter at least one, more often several, distinct calls and the males in addition produce more elaborate and longer lasting sound patterns which are called songs. Mountain bluebirds do not have an obvious song. Their nearest approach to it is a stringing together of a few of their usual throaty whistles which never carry far as they are generally uttered without opening the beak. Apart from the whistle, the only sound I have heard from these birds is a rapid clicking they made while hovering very close to me when I was checking the contents of their nest box.

The nesting of bluebirds at our acreage does not always go as smoothly as I indicated. This year our breeding birds were involved in emergencies which kept us in suspense. At the beginning of May, we had a brief view of a male. By the seventh, a female had established herself by a nest box near our cottage and she sometimes entered the box. This was unusual, as it is generally the males which stake out the nesting territory, but our bird was evidently alone. A week later, she was still in the same area, still interested in the nest box but still without a mate. Could she possibly acquire one with the breeding season now well advanced? Our worries on her behalf were resolved the next day when she was joined by the most intensely coloured male we had ever seen. The birds evidently paired right away, for the male popped in and out of the nest box and the female was seen to carry small feathers to serve as nest lining. Two weeks later there were five bluebird eggs and everything seemed to be going well. Early in June, when I moved the nest box roof and looked in, there was the female sitting tight on her eggs, but a few days later she disappeared and was never seen again. The box now contained only three eggs and a dead youngster, so small it must have died almost immediately on hatching. The male bluebird was still around. He often peered into the nest box and also into some of the holes of the martin house which was not occupied this year. More than before, he uttered his subdued song, perhaps in an attempt to attract another mate.

One can only speculate about the fate which had overtaken the female, but it is quite possible that she had been taken by one of the pair of Cooper's hawks which nested half a mile away. The hunting flight of one of these hawks sometimes took it through the very stand of trees which contained the bluebird nest box. The widowed male bluebird stayed in his territory for a month after the disappearance of his mate, but one day near the middle of July, he too was gone for good.

Just about the time one begins to see the early migrant birds, the "gophers" which until then have been hibernating underground begin to appear on the fields. It is probably no coincidence that the first red-tailed hawks, which prey heavily on the ground squirrels, show up at about the time the latter begin to appear above the ground.

During April the other species of ducks which breed in our area, both the surface feeders and the diving ducks, are present in good numbers and make use of the belt of open water which by then has formed along the shore of lakes and sloughs. Only the blue winged teal are latecomers not usually seen until the middle of the month. Walking about our acreage from late April onward, I am sure to startle a few female ducks off their nests. They run off frantically with thrashing wings to the nearest water. If that is too far they fly off once they are clear of the undergrowth which often surrounds the nests. It is disappointing how many of these nests come to grief. When I look at them a week or so later many are deserted; broken egg shells may be left in or near the nest; more often the eggs have vanished completely. This must be blamed on such predators as skunks, coyotes and magpies. There are no crows on our place during the nesting season and I destroy every magpie nest I find. But this does not eliminate raids by magpies living on adjacent properties. Occasionally a duck's nest is lost in a less dramatic manner. A pintail nest I found at the very edge of a pothole pond was flooded out after several days of heavy rain. The level of the small pond had risen by about two inches. Some ducks, nevertheless, succeed in hatching their eggs and about the beginning of June can be seen cautiously leading their string of downy yellow and dark brown ducklings to water.

One year a shoveller duck nested on the pasture near our dugout pond. When on the nest she was completely hidden by the surrounding grass but she was in full view **Shoveller Duck** from our cottage whenever some disturbance flushed her off the nest. She would walk back to it later. She always did this with great hesitation, generally stopping and sitting down briefly once or twice before completing the short trip. The horses on their way to water at the pond passed right by the nest, flushing the duck whenever one of them came within four feet or so. The duck then had to wait until the horses had watered and left the vicinity before she could return to the eggs. It often looked as if a horse were walking right over the nest and we feared that it must sooner or later be destroyed. But this did not happen, nor was the nest doomed by the fact that the horses cropped the grass around it down to a few inches. I happened to check this nest the very day the young had hatched. They were lying in the nest still wet from the eggs. A few hours later the duck led them down to the pond.

Well before ducks start nesting other summer birds such as

robins, purple finches, and several species of our native sparrows have also arrived. The earliest of these native sparrows to show itself is the tree sparrow. It only passes through our area to nest in the boreal forests of the north, mainly in the region where the forest meets the tundra. Another spring and fall transient is the water pipit, a greyish brown song bird whose "sip sip" call is easily recognised. The water pipits that pass through our district are on their way to the tundra, but others of the species nest on alpine meadows above the tree line in the Rockies. These high meadows resemble the tundra in many ways. In Europe this bird also nests in two areas which are equally far apart; on the tundras of the most northern parts of the continent and again above the tree line in the Alps.

Water Pipit

From late spring onward when the grass is already high, another native sparrow makes itself heard. Its song is weak, so weak that some people just cannot hear it. It sounds like the buzzing of an insect. Though extremely simple, this "tss bzzz" song is easily recognized once one is familiar with it. The singer is Le Conte's sparrow. Audubon, who first described this little bird, named it in honour of his friend Dr. Le Conte, a physicist and chemist. It is fairly easy to detect Le Conte's sparrows from their songs but more difficult to see them. Walking through the grass towards the source of the sound merely results in a sudden brief glimpse of a small bird flying up and away and then dropping down to disappear in the grass thirty or forty yards away. Patient watching while standing still is more successful. Sooner or later the little singer will become visible above the grass forest, low down on a bush, on a fence wire or on a tall tuft of grass. Le Conte's sparrow boasts no bright colours but presents an attractive appearance with buffy-orange and dark grey parts. A central white stripe flanked on each side by dark stripes on its crown helps to identify this sparrow.

Le Conte's Sparrow

The most striking thing about this bird is that it sings not only in the daylight hours, generally from the early afternoon onwards, but right through the darkest hours of the night. Whenever I have stepped outside the front door of our cottage during the night in June and July I have heard the buzz of one or more of these birds. None of the other related species of native sparrows, one of them also found in our region and three belonging only to the States, show this habit. The nightingale of Europe, named for its nocturnal singing, is not so marked a night singer as our Le Conte's sparrow for the nightingale sings in the daytime more than at night. The only suggestion to explain the Le Conte's night singing that oc-

Young sandhill crane
Photo by William Rowan

curs to me is that at night when other birds are silent the weak buzz of the Le Conte's sparrow more readily attracts females of the species.

Every spring on one or more days late in April or early May we hear the resounding "orr orr" calls of flocks of sandhill cranes

Sandhill Crane migrating northward high over our cottage. These birds stand about three feet high and have a wing span of almost six feet. Their voices carry so far that one may often hear them clearly and yet have to search the sky to see the birds. Sandhill cranes generally fly in a perfect V formation or in a line which is really half a V. With their long necks stretched forward and almost as long legs trailing behind they are unmistakable. Only the still larger and much rarer whooping cranes present a similar outline in flight. But whooping cranes are white with black wing tips while sandhill cranes are grey.

As the sandhills come overhead it is striking how rarely they give a wing beat. For long spells they glide forward on set wings as if propelled by some invisible force. Perhaps they are carried by a wind not noticeable at ground level, or maybe they gain height in the lengthy periods of circling which often interrupt their onward flight and then imperceptibly glide downward as they resume their northward movement. Two flocks often join with much calling and form a mass of birds wheeling about in the sky. Eventually they will drift on as a single larger formation or re-form into their original units.

On calm, sunny days we often see flocks of cranes pass over at intervals until the total reaches over a thousand birds. One morning early in May many years ago I was planting a number of small spruce trees on our acreage. I had no sooner started when the trumpeting call of cranes rang out and soon flock after flock flew overhead. By the time I had finished my chore near noon, I must have seen at least 2,000 cranes. When they had passed some distance north of me and must have been over Hastings Lake they started milling and circling about, but I saw no evidence that they actually made a landing. Indeed the observations of others indicate that migrating cranes only come down to the ground for the night and feed for a while in the early morning hours before resuming their journey.

My sighting of over 2,000 cranes in the course of a few hours is by no means a record for this general area. At Beaverhill Lake, a large shallow lake east of the Hastings Lake district, Mr. William B. Koski reported a huge movement of migrating sandhill cranes on April 26, 1981 in the magazine Blue Jay (Vol. 39, p. 197, Sept. 1981). In the initial wave, which passed over in three quarters of an hour he counted 19,000 cranes. Later many others which he could hear but not see passed over above the clouds so that it seems likely that altogether 30-40,000 cranes flew over the lake that day.

We know where these great numbers of cranes go to breed. Two subspecies or races of the sandhill crane nest in Canada. Those of the larger, more southern race breed in marshes and muskegs in British Columbia and the northern parts of the prairie provinces. The advance of agriculture has destroyed many of the former breeding haunts of these birds but their population in Canada must still number a good many thousands.

The smaller race of the sandhill crane, or little brown crane, breeds over a vast area from northeastern Siberia and northern Alaska across arctic and subarctic Canada. There are probably still several million of these birds and no doubt they form the majority of those which migrate over our district. But the two subspecies cannot be told apart in the field. The cranes we see in the spring have come from wintering grounds in California and some of the southern states. Others are from even more distant areas in Mexico and Cuba.

I have seen sandhill cranes of both these subspecies on their breeding grounds. The first occasion was in May, over thirty years ago, when the late Professor William Rowan took me for a four-

Greater yellowlegs on nest
Photo by William Rowan

day trip into the wild country which lies between the lower course of the Pembina and the Athabasca river. There beyond the last farms lies a huge muskeg where sandy ridges covered with jack pines separate muskeg tracts scattered with black spruce. Some of the wet areas have small or fair-sized lakes. On each of the larger lakes there is generally a pair of loons. Walking is tricky except on the jack pine ridges for there are patches where the foot conveys the sensation that the wobbling mat of vegetation which supports it is afloat. The wanderer feels he may at any moment break through into ice-cold water of unknown depth.

These muskegs were home for some interesting birds such as dowitchers and the two kinds of yellowlegs. The relatively-rare greater yellowlegs which nested on the san-

Greater Yellowlegs

dy ridges had been of particular interest to Rowan. In earlier days when he supported a large family on a meagre salary, the price their eggs commanded from collectors came in handy. Rowan was almost always collecting when in the field though at this time he did so for the zoology museum at the university rather than for himself. He had just shot at (but missed) a pair of ring-necked ducks when we heard the faint "rorr" call of a crane in the distance. We stopped to listen and the call was repeated a little louder, then louder yet. Soon it sounded quite close. Bending low, we walked toward the apparent source of the sound. For some time we found nothing and the calls ceased. Yet we knew the bird must still be very near for we would inevitably have seen it above the intervening vegetation if it had taken wing. At last through a gap in some shrubs we could see the crane walking to and fro like a sentinel, its drooped wings almost touching the ground. Instead of being slate grey as the books depict it, this bird like others I saw later had rusty brown colouring. Rowan explained that this is due to deposits of iron oxides on the feathers of cranes which feed a great deal in waters rich in iron. The rusty discolouration of the head often seen in snow geese and whistling swans arises in the same way.

The crane we were watching was soon joined by its partner. We felt sure that their nest must be near at hand so we advanced into the open to look for it. The cranes, of course, flew up trumpeting loudly and soon were out of sight. But our search for a nest was in vain. Still curious, we concealed ourselves at the base of the nearest sandy ridge where some small spruce and shrubs provided good cover. The sky was grey and a stiff breeze was blowing. After we had been there half an hour the cranes silently planed in and landed. For a long time we watched them appearing and disappearing behind trees or shrubs in the middle distance. Though we never saw one of them sit down as on a nest we felt we had pinpointed the area they were most interested in. Inevitably disturbing the cranes again, we searched once more and this time Rowan found the nest. It was merely a flattened heap of marsh plants and roots about three feet across surrounded by the open bog. A small lake was nearby. The nest held two large, pale-brown eggs with markings of darker brown.

It had started to rain by the time we found the nest and we decided to head back to our temporary home in the village of Fawcett as fast as the terrain would permit. The rain soon became a downpour and lightening and thunder were added to the action. I was wearing shorts and soon realised, as my legs got slapped by soaking twigs and scratched by snags, that they were not designed for this sort of country. Nevertheless, by the time we reached the edge of the cultivated land and the rain had eased off, I felt I could look back with satisfaction on my first experience of the muskeg.

The following day we walked further into another part of the muskeg. As we were returning we suddenly flushed another sand-hill crane from a nest with two eggs. The bird had evidently been incubating and we had taken it by surprise. After jumping up from the eggs it walked off in a sort of crouch with lowered head and neck and drooping wings. When it was some little distance away it betrayed its emotional state by pulling up and flinging beakfuls of moss. We moved on as soon as we had taken a quick look at the nest and so gave the crane a chance to calm down and return to its eggs.

Some days later we revisited the owners of the crane's nest we had found the first day. Now that we knew exactly where it was we were able to see the bird on the nest before it took off. All one could make out was a rusty brown heap. This was the crane's body. On spotting us it had laid its head and neck flat along the ground where the low vegetation hid it completely. At our closer approach it stood up and strutted about with partly extended wings until we were once more out of sight.

The following year, again about the middle of May, I revisited the same muskeg area with my wife and two-year-old son. When Rowan and I had gone there we had crossed the Pembina by ferry. But progress had now reached this backwater and a new bridge spanned the river. It was a warm, sunny day and we settled down for a picnic by the shore of a lake not far from the edge of the muskeg. We had been sitting for about half an hour when I was suddenly aware of a bright red spot on a marshy peninsula which projected from the opposite shore of the lake. Binoculars revealed that the red spot was the patch of bare skin on the forehead of a sandhill crane. This bare area is a normal feature of the head of this and some other kinds of cranes. The one I was looking at was lying flattened on its nest just like the one we had seen the year before. Its head was pointing straight at us. Evidently this is how sandhill cranes on the nest conceal themselves from potential enemies until more closely approached.

Two years later I spent the summer in the Arctic on Banks Island. It is one of the many breeding places of the little brown crane in the far north. The small settlement where I was staying with an eskimo family lies about twelve hundred miles to the north of the Alberta muskeg where I had seen my first crane nests. It was not surprising to find that when the Alberta cranes already had eggs, their Banks Island relatives were only just arriving from the south. They flew in over the frozen sea when the land was still largely covered with snow. It must be difficult for the cranes to find food at this season. Indeed, the eskimos, who hunted them to some extent, said the cranes are in good condition on arrival but then start losing weight. Only when most of the snow is gone and they can get at the spring vegetation do they regain their earlier weight. For the first two weeks after their arrival the cranes I saw were either searching for food or merely standing about and looking rather sad. Not long after, on the second of June, I saw a pair perform one of the dances which are a feature of several kinds of cranes. One of the pair suddenly jumped up and over its partner landing beyond it. There he bowed, almost touching his head to the ground. Then, in alternation with the other bird, he would point his beak at the sky and leap three or four feet up into the air while both birds called loudly. A series of such bouts of jumping, with spells of bowing with dropped wings, made quite a performance.

Though I saw this particular dance in the spring it is not a courtship performance connected with pair formation. These dances may be performed at all times of the year, and cranes mate for life. When cranes dance they seem to be expressing feelings of well-being or even exuberance. If you like they're just letting off steam. The mutual performance of these rituals probably serves to strengthen the bond between the members of a pair.

It was not until the fifteenth of June that I found a crane's nest on Banks Island. It was on a hillock on a gently sloping tundra hillside and held the usual two eggs. A number of other interesting arctic birds were nesting nearby. There were Brant and snow geese and a pair of glaucous gulls. When I checked the crane's nest three days later it was empty but the broken egg shells were lying near by. I blamed the glaucous gulls. These large birds are determined predators of eggs as well as young birds. On the other hand, an arctic fox might have raided the nest during some brief spell when neither of the cranes was close by. Other cranes in the area did, however, breed sucessfully. By the middle of July I saw them with half grown young.

In September flocks of cranes, now on their southward migration, appear once more in the skies over central Alberta. Sometimes a flock interrupts its direct flight and comes down to a somewhat lower level. The birds then break formation to wheel about in a leisurely manner. One can then make out a new call. In addition to the resonant "orr orr" call, the only one heard from the spring flocks, there are also higher pitched calls that sound like "peeri". These must come from the young of the year which, though fully grown, have not quite acquired the powerful voice of the parents.

Spring is the season when two-year-old beavers are expelled by their elders from the lodges in which they were born. These exiles

Beaver

wander about looking for a piece of lake shore, a stretch of creek, or a pond where they can set up their own home site. These searches are carried out at night for one virtually never sees a beaver any distance from

Beaver workings

water. It is reported that it is always a female which sets up a new site. When the fall mating season approaches she is likely to be joined by a young male and to deliver her first young the following spring.

The waters on our acreage are so shallow that when they freeze there is not enough depth beneath the ice for a beaver to keep its store of winter food. Apparently this is not obvious to every wandering beaver. So it was that one May morning we were surprised to see a beaver on the edge of our small marsh. When full it is hardly more than 18 inches deep. Even on land, beavers generally look dark for their fur is still wet from their latest swim but this animal must have been out of water for quite a while. Its fur had dried in the sun to an attractive reddish-brown tinge. Perhaps it had only arrived early that morning and was still resting when we saw it. It stayed about the marsh for eight weeks. We had proof of its continued presence from the increasing number of young poplars it cut down. Among a pile of old fallen sticks and branches very near the spot where we had first seen it, this beaver made a large hole which may have had an under-water exit. Had the animal stayed this might have become the foundation of its lodge. It left with the advance of summer when the marsh became so shallow that it, too, noticed how unsuitable its new home really was.

Half a mile from our place a lightly paved secondary road runs

Flying squirrel

along the shore of a fair sized pond. A pile of poplar branches and sticks partly in the water and partly on the strip of grass between the road and the pond showed that some beavers had made a burrow in the bank. One summer day when we drove by the pond the road surface fell in behind the rear wheels of our car, leaving a gaping hole into the excavation the animals had made under the road. Fortunately they had thinned out the roof of their burrow only to such a degree that it gave way very slowly under pressure. Soon after this the hole was filled by our municipal roadmen. The beavers accepted the eviction from their home in the bank, built up a little island fifteen yards from the shore and erected a fine new lodge in there.

One sunny afternoon about the middle of June, I set about destroying a magpie's nest in a poplar wood on our acreage. From

Flying Squirrel the top of our rickety ladder I could just reach the nest with a long pole. A series of upward jabs caused the sticks and clay which composed the nest to come hurtling down, some of them on top of my head. I had hardly started the attack when, not surprisingly, a magpie flew out of the nest and disappeared silently. But I did not expect a flying squirrel which minutes later emerged from the lowest part of the nest. Here it too had its own summer home. The squirrel jumped from the nest across to the trunk of the nearest poplar. Though these attractive rodents cannot fly in the true sense of the word they can glide for impressive distances. They stretch out the loose skin between wrist and ankle on each side of the body. Their long, wide tail also adds gliding surface and stability. I had never seen one of these glides so I shook the tree on which the squirrel had landed hoping this would persuade it to launch itself once more. In a smooth curve it swooped off its perch twenty-five feet up; descended to within four feet of the ground thirty yards away from its starting point; and then managed a final short upward glide to land about six feet up a tree trunk. Just then one of my sons came along and in order that he too might see what this remarkable little animal could do we induced it to show us another glide. This time it landed on one of the trunks supporting an old tree house my sons had made in their boyhood. It climbed up to that half ruined structure and vanished inside. We never saw it again.

During the early hours of a bright spring morning, the air about the acreage is filled with a variety of bird songs and calls. There are **Purple Finch** always a few robins loudly carolling away with varied short phrases which they are apt to string together for quite some time. A rapid, lively and melodious warble is delivered from its spruce perch by a purple finch. His female, an inconspicuous grey bird, may be sitting on the nest in the same tree. There was only one spruce on our land when we bought it twenty-four years ago but many I planted are now twenty or more feet tall. They have attracted not only purple finches but chipping sparrows. Neither of these species were seen on the place before the conifers were planted.

As a singer the chipping sparrow comes very low on the ladder of bird musicians, for its song is merely a fast, long-drawn **Sparrow** monotonous trill, which remains on the same pitch from beginning to end. The clay coloured sparrow, a common bird in bushy areas where woods and meadow meet, sings no better for it merely utters three or four low-pitched buzzing sounds. The song sparrow which lives in similar habitats is, as its name suggests, a better performer for it uses musical as well as somewhat buzzy notes. Its song usually starts with a repetition of three or four notes that sound something like "pweer pweer pweer" followed by a few fast sounds of descending pitch. The song as a whole can be rendered in words as "prees prees presbyterian". The somewhat larger white-throated sparrow keeps more to the woods. It sings in clear notes which sound just like a slow whistled rendition of "hard times Canada."

Out on the pastures, often delivered from fence posts, the thin "tsi tsi tsi tsee tsay" song of the Savannah sparrow is frequently heard. From time to time a goldfinch will add its light canary-like song to the general chorus.

Back near the cottage the house wren is a persistent singer. It starts off with an explosive burst and follows this with a hurried **House Wren** jumble of notes which fall off in pitch toward the end. Though we have set up two nest boxes with particularly small entrance holes for these birds, one pair prefers to build a messy-looking nest of small twigs under the eaves of the cottage. The busy comings and goings of the parents are often interupted by bursts of song when they bring food for their nestlings.

Only two species of warblers nest in any numbers about Hastings Lake, although members of several other species pass **Yellow** through in late May and again in August and September. Our two breeding warblers **Warbler** are the yellow warbler and the yellow-throat. Like most wood warblers, the group to which North American native warblers belong, they are showy birds. The male yellow warbler is bright yellow below with orange streaks on the breast. But his song, a cheerful sounding "tsee tsee tsee tsee ti wee", could hardly be simpler.

The male yellowthroat has a black mask lined with white above a bright yellow throat. It presents a striking appearance but rather **Yellowthroat** monotonously sings nothing but "witchity, witchity, witchity". Comparing our wood **Warbler** warblers with the Old World warblers which include outstanding singers like the nightingale one feels that the wood warblers are very attractive to the eye but can barely sing. Old World warblers sing beautifully but with their simple brown and grey colours are not much to look at.

Some of our best songsters here are the vireos, woodland birds of simple colouration. The red-eyed vireo loudly flings out short **Vireo** robin-like phrases which, after a pause between each phrase, it repeats again and again. The warbling vireo, a slightly smaller bird, delivers less loudly a more continuous pleasant warble. It is more like the song of some of the Old World warblers than any other bird I have heard in North America. The bird is common in the poplar woods on and around our acreage.

The swallows may not be particularly thought of in connection with song. Instead, tree swallows, several pairs of which occupy our nest boxes, produce little more than a twitter. The calls and song of the barn swallow are much more appealing. As they swoop over the fields their simple "kweet" always sounds so cheer-

Barn Swallow ful and so does their long musical twittering song in which occasional guttural sounds are interspersed. Barn swallows are as common in the European countryside as they are here. I remember them vividly from my early boyhood in Switzerland and from later years in England. Wherever they occur they build their mud nests onto the walls of farm buildings or onto horizontal beams which are sheltered from above in sheds or barns. Farm buildings attract them not only because they offer nest sites but because the livestock and its manure attract insects which the agile swallows snatch up on the wing. The many centuries during which these birds have been closely associated with farming make them a veritable emblem of rural peace. In primitive times farm swallows no doubt nested in caves or on cliffs provided with sheltering overhangs. They were probably much less common before the advent of agriculture. Many pairs of these birds breed in my neighbour's pigbarn. A few even nest on his house. I would dearly like to have just one pair on my place. To entice them I built a small sloping roof, with a shelf as a nest site underneath, first on the south wall and next year onto the east wall of our small stable but no swallow became interested. Then I left the windows of the stable permanently open. One year a pair started to build on a beam within but they gave up long before completing the nest.

Most conspicuous among the birds that spend the summer in our woods are the northern orioles. Between seven and eight

Northern Oriole inches long they can be considered medium-sized birds. The strikingly coloured black, orange and yellow males are unmistakable. Their songs with their full toned whistling or piping notes, which include such phrases as "peooeeoo" are equally characteristic.

Though less noticeable than the arrival of migrant birds the appearance of garter snakes about the first third of April is just as

Garter Snake definite a sign of spring. They are the only kind of snake found this far north in Alberta and they are not poisonous. Not only are they harmless but they are rather beautiful. Behind the dainty head an orange stripe runs along the back and there are yellow stripes running the length of the otherwise black body. They feed on small frogs, particularly the tiny western chorus frog or "spring peeper". This is the frog which for two or three weeks from mid-April, keeps up a continuous chirping in our ponds and sloughs. Garter snakes also eat a variety of insects. Like most snakes they can open their mouths far wider than seems possible and so swallow, with many repeated small gulps, surprisingly large prey.

One morning I came across a three-foot garter snake which had taken on more than it could chew. It was lying on a bed of floating algae at the side of a slough and from its open mouth there projected all but the head and neck of a naked young redwinged blackbird which was certainly larger than a full grown sparrow. It had probably fallen from the nest and the snake in the usual manner of its kind had swallowed the bird's head, but could not get it down any further because the wings projecting on each side made a barrier much too wide to be engulfed.

Through the summer months only single garter snakes are occasionally encountered but on sunny days early in October we are apt to see several, all crossing the road from north to south. Quite a few of them meet their death under the tires of passing cars. These snakes are on their way to a wintering place on a farm a mile and a half to the south. Here they hibernate in underground holes just below the frost line, clinging together in dense tangles which helps to minimize their heat loss. The owner of this farm has told me that in spring they emerge in hundreds from their winter quarters. Studies at garter snake wintering places in limestone caves in Manitoba have shown that at this season males wait around the exit holes to fertilise the females as they come out into the open. Unlike many other kinds of snakes female garter snakes give birth to live young some three months after they have been fertilised. Up to 30 young may be born to one female, usually late in August. The young have to fend for themselves as soon as they leave their mother's body and later, in some as yet unknown way, find their way to the nearest hibernating place.

There is a cellar, simply dug out of the ground, underneath our cottage, where we often keep cartons of beer and cider. It can be

Tiger Salamander

reached by means of a trap door in the kitchen floor. One spring afternoon my wife was about to get something from the cellar but no sooner had she raised the door when she shrieked "there's a dragon there!" and slammed down the trap door. Curious to see what had alarmed her and knowing no really alarming beast could be sheltered there, I armed myself with a flash light and lifted the trap door. There was the monster, ugly indeed, with its large flattened head and wide mouth; but only about ten inches long. It was a tiger salamander, a long-tailed green amphibian with dark blotches. The tiger in its name comes from the eastern representatives of the species which have yellow blotches on a dark greenish-brown background. Our western ones merely show the dull colours mentioned. Having identified the harmless creature I simply closed the trap door. He was to remain the last adult of his kind we were to see. Next summer, but only during that one summer, our dugout pond was simply swarming with salamander larvae. These are young salamanders which correspond to the tadpoles of frogs. Adult salamanders live on land except when they enter water to breed, and they breathe by means of lungs. The larvae have feathery gills by means of which they breathe in the water. When they reach the right age in late summer the gills shrink away and the animals become air breathing and leave the ponds in which they hatched.

Somehow my spring account has drifted from vocalisations of our most prominent song birds to the lowly and silent snakes and **Common Snipe** salamanders. To return to birds, one of the striking sounds of spring in our area is the winnowing or drumming of the common snipe. It is performed by males flying at sixty or more feet. From this height they suddenly dive toward the ground at an angle of 45 degrees and produce a tremulous humming "oo oo oo" sound for 2 or 3 seconds. It is a performance that is repeated over and over and it is evidently their equivalent of a song. It is produced by the vibration of the bird's outer tail feathers caused by its fast descent. The tremulous character of the sound is due to a slower quivering of the half open wings. The basic hum is easily imitated by fixing the two outer tail feathers to a cork attached to a string and whirling it around swiftly.

Common snipe live not only in North America but also in Europe, much of Asia and a large part of Africa. I had long been familiar with snipe in England and when hearing the American birds drumming for the first time was immediately struck by the fact that their hum was of a higher pitch than that which I remembered from England. This seems to be explained by the markedly narrower outer tail feathers of American as compared to Old World common snipe. Apart from their "instrumental" song, snipe use their voice to utter a characteristic spring call which resembles the words "jick jack, jick jack", while at all seasons they flush with a harsh scraping "scaap" call. Though their nests are always on marshy ground, drumming snipe cruise the skies quite some distance from their homes.

Near any cattail marsh one can hardly escape hearing the song of red-winged blackbirds. It is a loud "konk a reee". The last por-

Red-winged Blackbird tion of this song sounds so mechanical one is reminded of the ring of a telephone. While singing the birds, which are otherwise black, raise a patch of bright red feathers on their shoulders and create a very striking sight. Every spring they arrive about ten days before the slightly smaller brown-stripped females.

Singing "redwings" are easy to spot but another marsh dweller is heard far more often than it is seen. This is the secretive sora rail.

Sora Rail Its "song" is a fairly musical whinny which descends in pitch as it goes along. At times it also utters an emphatic "pwooeep" or a sharp "keck". Occasionally when wading through the sedges on the edge of a marsh one suddenly flushes one of these nine inch rails. It appears for seconds as a brownish shape with dangling legs. After a short flight it drops down and vanishes among the tangled vegetation. Some patience is required to get a good view of one of these birds. The best way is to watch a pool in the part of a marsh from which calls have recently been heard. Sooner or later a rail will emerge from the dense mass of water plants and walk along its edge. Binoculars will reveal its short yellow beak and long legs of the same colour, the bands of dark grey and white on its side and a pointed cocked up tail which is twitched with each step. Several times I have waded about our marsh in hip waders looking for nests of the sora rail without finding one. Yet I know they breed there because early in July I have seen their chicks, little balls of black down, beside one of their parents.

Rather smaller than the redwing is its relative the brown-headed cowbird. Male cowbirds, apart from their heads, are glossy bluish **Brown-headed Cowbird** black while the females are a dingy brown colour all over. They are birds of farm land which habitually walk around grazing cattle or horses and pick up the insects disturbed by the movements of the animals. Before domestic livestock were brought to the west cowbirds congregated in the same way about the buffalo. Male cowbirds have a characteristic short bubbly song and seem to mate indiscriminately with any receptive female. The latter are peculiar, for they are our only brood parasites; birds that lay their eggs in the nests of other birds which hatch the foreign egg and feed the youngster which emerges from it. The young cowbird, being so much larger than its nest mates, generally pushes these out of the nest. It goes without saying that cowbirds do not take the trouble to build nests. Cowbird eggs do not particulary resemble the eggs of the many different kinds of small birds whose nests they parasitise. Yet in most cases the strange egg is accepted. This is not the case with the European cuckoo whose habits were already known to Shakespeare. Cuckoo females specialize, and lay their eggs in the nests of only one kind of host. The cuckoo eggs resemble those of their hosts in size and colour to a remarkable degree. Our cowbird simply lays what might be called a standard egg which must look strange to nearly all its hosts. Yet it is successful for it is certainly a common bird.

Even after young cowbirds have left the nest their foster parents look after them for about two weeks. I recall seeing dainty chipping sparrows on our driveway eagerly feeding their much larger cowbird foster children. On other occasions song sparrows or clay-cloured sparrows have been the foster parents. It is not only the brown-headed cowbird and the cuckoo of Europe which propogate their kind in this way at the expense of other birds. There are in fact nearly eighty brood parasitic bird species. With the exception of Antarctica no continent is without them. The brood parasitic habit evolved independently several different times but we still do not know why. Nor do we know how the physiology of the female parasite is modified to block its nest building and incubating urges.

Larger marshes or sloughs than those found near our acreage are the summer home of bitterns. These are large birds of the heron tribe which have a wing spread of 3 1/2 feet. In spite of their size they are rarely seen for they keep to the cover of tall reeds or cattails where they hunt frogs, small fish and, occasionally, garter snakes. It is their loud spring call or booming which most often indicates their presence. It is a sound like no other, a few hiccuping grunts followed by an emphatic "oom pa klook". From a distance only the thumping "pa klook" can be heard and this is what I heard from our cottage one evening late in May. I traced the call to a slough on my neighbour's place. Remaining in the cover of the belt of willows which surround the slough I could see the bittern standing at the waters' edge in front of the sedges of the far shore. What is more, I was able to watch it calling. A swallowing movement with lowered head accompanied each preliminary grunt, then it flung up its head in a gulping movement and uttered the loud "oom pa klook".

It had chosen too small a habitat. Repeated calls apparently failed to attract a mate and it soon left the area. Bitterns do, however, nest somewhere in the district. Nearly every summer single young birds, already able to fly, are seen.

Bittern

The Old World also has a bittern. Though a different species it is very similar in appearance to ours. Its boom, which I have often heard in the Norfolk Broods in southeast England, is quite different; after a few grunts there is a single "oom" reminiscent of cows' bellow. This bovine bellowing must have been the basis for the scientific name of the bitterns: *botaurus* — meaning oxbull.

Through the drainage of marshland bitterns became extinct in Britain in the middle eighteen hundreds. Early this century bitterns wandering across the sea from Holland re-established themselves. At present they are nesting again, in modest and protected numbers, in several parts of England.

The coot is a far more common marsh dweller than the bittern and one that makes no particular attempt to hide itself. About the

Coot

size of our smaller ducks it is black with a white shield extending up its forehead and a patch of white under the tail. Being a rail and not a duck its feet are not webbed. Instead it has rounded lobes of skin along the sides of its toes. Like webs, but with less efficiency, these increase the surface of the foot when it is thrust against the water in swimming. Coots quite often bring underwater vegetation to the surface where they pick out the tid bits that look most tempting to them. Our other rails, the sora, virginia and yellow rail, do not swim to any extent and don't dive at all. In consequence they have long, unlobed toes which enable them to wander over floating water plants. Soras are more carnivorous than coots, they take insects, small snails, and crustacea as well as seeds.

In Alberta the coot is popularly called the mudhen and it is despised as a target by all self-respecting duck hunters. It is, however, quite edible, for on the lake of Constance, where Switzerland, Germany and Austria meet, an annual coot shoot is held to harvest some of these birds which winter there in great numbers. Coots, though not as musical as soras, are noisy enough. Their most musical call is a repeated "ka ha, ka ha."

The marsh on our acreage was created by my older son and myself simply by damming the roadside ditch along our lowest lying pasture. Pools, fed by the run-off from adjacent higher ground, soon formed and, over the years, stands of cattails grew up. It was a great pleasure to us when this attracted soras, redwings and a variety of ducks. The coots, which also came in, gave us particular pleasure for their downies are not simply black all over like those of other rails, but have bright red beaks and orange down on the head, neck, back and wings.

No bird gives such obvious demonstrations of defence of its territory as the coot. In spring and early summer coots are almost always swimming about in a threat posture. Head and neck are lowered to the surface of the water and the tips of the closed wings are raised to give the body a humped appearance. Steaming along in this attitude, spitting his harsh calls, the aggressive coot chases others of its kind from the stretch of water and adjacent shoreline which constitute his and his mate's territory. Coots need not be concerned about ducks or grebes but, by their blustering behaviour, they often manage to scare these birds away as well.

The territory is the feeding area of the pair and will, in due course, hold their nest. When the male needs support the female coot also takes part in territorial defence. Sometimes two pairs of coots confront one another in threat posture, most probably on the boundary between two adjacent territories. Suddenly one attacks his opposite with jabbing beak, then throws himself on his back and kicks at his opponent with his large, clawed feet. The bird being attacked also lies back and strikes with his feet, flapping to keep afloat. The contestants may then seize one another's feet and continue the fight with beaks alone. Vigorous as these fights are they never last long. The bird which senses that it is getting the worst of the conflict withdraws before it is seriously injured. With very few exceptions this is true of all territorial fights and struggles connected with sexual or territorial rivalry of birds and mammals.

A very beautiful small grebe, the horned grebe, breeds on nearly every slough, even the small ones in our district. It is one of the

Horned Grebe
several water birds popularly known by the strange name of helldiver. The horned grebe's nest is built by both members of the pair. It consists of wet, often rotting, water plants built into a platform which is usually floating (though it is kept in place by adjacent still-growing water plants). Male and female are equally colourful with large orange ear tufts above black cheeks, an orange-brown neck and breast and a black back. These grebes are not only common all over the west but are also found in northern Europe and northern Asia. We take them for granted but in Britain only a few scattered pairs breed in Scotland. They are so rare there that an ornithologist friend of mine who had marked on his maps the waters where he had found them nesting, was worried that if he ever lost his maps they might get into the hands of egg collectors who would rob the birds. The downies of all grebes have a striking pattern of black and pale-brown stripes. Their parents have the attractive habit of letting the young climb on to their backs where the chicks find warmth by burrowing under the parent's wing.

The pied-billed grebe, appropriately named for its beak, is in fact pale-grey and black. It is less colourful than the horned. Apart

Pied-Billed Grebe

from some black markings about the head, it is brown. It lives on the same sort of waters as are used by the horned grebe but is less common.

We have yet another small grebe. Like the two just named it is about 9 inches long, and somewhat resembles the horned except

Eared Grebe

for an upturned bill and black neck. This is the eared grebe, named after a tuft of yellow feathers which cover its ear region. It breeds in colonies on some of our lakes and large sloughs. In late summer grebes leave their young and begin to moult into their winter plumages. Both horned and eared grebes become dark grey above and white below; they are then difficult to tell apart. These grebes are quite silent in winter and do not call loudly even in the nesting seasons.

Our last grebe, the 14-inch red-necked, is handsomely marked with a black cap, white throat and orange neck. It is often quite

Red-necked Grebe

noisy, uttering a long-drawn whinnying or a braying "a a a a a a" which can be heard from afar. It settles on larger waters than the smaller grebes. On Hastings Lake two pairs often breed in a bay where a road runs along the lake shore so that one can easily see the birds on their nests from the car.

The great blue heron, even taller than its relative the bittern, is an occasional visitor to most of our sloughs and marshes. In our

Great Blue Heron

immediate area I know of only one breeding colony of these birds. They nest in trees on an island in a medium-sized lake. As there are cottages on another part of the same lake one wonders whether disturbances may not eventually force the herons to leave. However, they also nest in Elk Island National Park where they are safe from disturbance.

A considerable variety of ducks nest in and about our sloughs and marshes. The most peculiar and colourful of these is the ruddy

Ruddy Duck

duck, easily identified by the characteristic way it holds its tail almost upright. Ruddies of both sexes can also be told by their white cheeks but the male also has a bright blue beak, a black cap above its white cheeks and a rich, reddish-brown body colour. In spring these little diving ducks are not at all shy and the startling courtship of the males is easily observed. They swim about with tail tilted forward so that it almost touches the back of the head. The head is raised and then drawn down on the puffed out chest spasmodically while a wooden sounding "chuk uk uk uw" is sounded. This performance is repeated so jerkily one is reminded of a mechanical toy.

The displaying drake generally erects a patch of the black feathers above each eye forming two protuberances like the knobs from which deer grow their antlers. Seen from the front these bulges look very odd and to my mind give the bird a decidedly menacing appearance. A drake may display at a duck, or at other males which it often attacks in a rush when they come displaying too close to the duck on which it has made a claim. So strong is the urge to perform this display that lone males sometimes go through it when no other duck is in sight.

In addition to the snipe, three other species of small shorebirds nest in our marshes and sloughs. They are the lesser yellowlegs, the spotted sandpiper and the killdeer plover. At the larger lakes such as Beaverhill, Miquelon and Joseph lakes there are, in addition, three larger nesting shorebirds; the marbled godwit, the willet and the avocet.

Killdeer often nest some distance from water and their dark blotched eggs, laid in a shallow scrape on the ground, blend so **Killdeer** well with their surroundings that they are difficult to find. When disturbed at the nest the plover runs off in a crouch with fanned-out orange tail (its most colourful part), waving its wings as if they were injured and uttering piteous cries. The usual reaction to this "broken wing trick" by the human observer who merely wanted to see the nest is to leave in order that the distressed bird can calm down. Predators, however, follow the apparently injured bird until they have been decoyed to a safe distance from the nest. The bird then escapes them at the last moment. Not only killdeer but a number of other ground-nesting birds draw their enemies away from their eggs or young in this manner. Like the snipe, the killdeer engages in song flights but his music is vocal, not instrumental, and consists of long-drawn repetitions of its name. Though strictly a North American native, the killdeer occasionally turns up as a vagrant in Britain and is included in some British bird books. I well remember seeing a colour picture of a killdeer in such a book many years ago. I admired the bird's black double neck band and its orange rump and tail and felt convinced I would never in my life see anything so colourful.

The lesser yellowlegs is so named to distinguish it from its larger relative the greater yellowlegs which only passes through our area to nest in the muskegs of the forests to the north. It is a dainty, long-beaked and long-legged bird, mottled brown in colour except for its white rump which only becomes obvious when it flies up. It **Lesser Yellowlegs** too has a song flight and its pleasant repeated "pee a wee" song often provides the first sign of its spring arrival. Yellowlegs, from nests I have never yet been able to find, bring their young to my neighbour's slough every year. Though the young have so far remained invisible to me in the lush vegetation, I can tell from the frantic behaviour of the parent when any one passes along the adjacent road that the young must be nearby. The old birds perch on a fence post or on a power pole continually calling "tyoo, tyoo, tyoo" and take flights from one perch to another, only stopping their frantic calling when the danger is passed. Often in their excitement they perch on the slender power lines teetering to and fro with waving wings and only just managing to keep their balance.

Every spring and summer there are some spotted sandpipers about our dugout pond and our neighbour's slough. They are **Spotted Sandpiper** rather inconspicuous birds just over 6 inches long, grey above with rounded black spots below and rather short beaks and legs. They trip along muddy shores with head and body tilted downward and continuously bob their rear ends up and down. When they fly the wings are slanted down below the horizontal and only their outer parts make rapid flickering beats. As they go they utter a rapid "peet peet peet peet peet". Nearly every year we find the four eggs of these birds in a simple grass nest. These sandpipers are known to defend their nesting territories against others of their kind and both male and female take part in the encounters.

About 5 o'clock on a June morning I saw two pairs of these sandpipers on the shore of our dugout. One bird repeatedly strutted in a very imposing posture after one of the others, in a display probably meant to intimidate the other bird. It stood upright, bill directed forward, and held its half open wings in the vertical plane. Seen through binoculars it looked quite menacing and reminiscent of a heraldic eagle. In the intervals between this posturing there were vigorous fights, two birds darting at one another with fluttering wings, each stabbing at the other with its beak. The whole performance was evidently a dispute over territory.

Recent studies on individually marked spotted sandpipers have revealed that their sex life is quite unusual. When a female has laid the third of her clutch of four eggs, she often prospects for a new mate. After laying the fourth egg of her first clutch, she lays a sec-

Female Wilson's phalarope in breeding plumage

ond clutch for her second mate and goes on to a third male for whom she also lays a set of eggs. The males take over the eggs, incubate them and look after the chicks that eventually hatch. The advantage of this method of reproduction is believed to be that it increases the likelihood of survival where predation on eggs or young is heavy. A female, instead of laying only one clutch per season which might well be discovered by a predator, lays several. At least one or two are likely to remain unharmed.

In late June or early July I often come across spotted sandpipers which by their anxious behaviour and continual calls show that they have young nearby. Since reading the study mentioned above I know the agitated birds are most probably males. I have a simple method of getting a view of their chicks if the adult, as is often the case, is stationed along a road. I drive the car to within about 60 yards of the parent sandpiper and watch. The stationary car does not frighten it and very soon it stops its alarm calls and in no time the dainty long-legged downies come out of the roadside vegetation into full view.

Some of our largest sloughs are the breeding places of another somewhat larger shore bird, Wilson's phalarope. The word

Wilson's Phalarope

"phalarope" comes from the Greek and means "coot-footed". Phalaropes have little lobes at the sides of their toes like coots (*phalaros* in Greek). This is related to the fact that they are the only shore birds which not only wade in shallow water but swim quite freely. Apart from Wilson's there are two other kinds of phalaropes which nest in the subarctic and arctic. These winter at sea, feeding on the marine plankton and resting afloat.

Family life among the phalaropes is even more peculiar than that of the spotted sandpiper. While the sexes look alike in winter, the female of all the three phalarope species has a far more colourful breeding plumage than the male. A female Wilson's phalarope is in fact most elegantly coloured. Below a pearl-grey forehead and crown she has a white spot above the eye. A dark stripe runs through the eye and becomes a rich cinnamon colour as it passes down the side of the neck and runs over the grey back. The underparts are white. A long fine beak adds to the daintiness of the whole appearance. The male has only a dingy version of the head and neck pattern and a brown back. In the spring when phalaropes often swim about in small flocks, a female generally follows a particular male and keeps other females away from him by threat postures or actual attacks. The males at first show objection. to the nearness of the females which have selected them, but after a while seem to get used to their presence. In this way, entirely as a result of female initiative, pairs are formed. When the female has laid her four eggs in the grass beyond the marsh vegetation of slough or lakeshore she takes no further interest in them. It is the male, in all 3 species of phalarope, who does all the incubation and after three weeks hatches out the chicks. He then guides and guards them till they are old enough to fend for themselvs. The females, who have completed their duties once their eggs are laid, gather in small flocks at certain of our lakes and soon disappear on a migration to winter quarters in South American marshes and lakes.

When male and female birds of the same species differ in their plumage, the male in the great majority of cases is more colourful than his partner. This is the case in peacocks, pheasants, most species of ducks, the red-winged blackbird mentioned above and many others. The phalaropes with their more colourful females are therefore very unusual. This coupled with the female's aggressive behaviour and the fact that only males look after the eggs and young presents intriguing problems. Since hormones influence feather colour and behaviour in certain birds I undertook a study to determine whether the unusual features of phalaropes could be explained on the basis of their hormone production. A number of the birds were collected under special license and in the laboratory I measured the male and female sex hormones in their sex glands as

Wilson's phalarope in winter plumage

well as one of the hormones of their pituitary gland called prolactin. From studies in other birds prolactin was known to be connected with the urge to incubate eggs. An explanation of the fact that only male phalaropes incubate emerged when I found in females only a fraction of the amount of prolactin present in male pituitaries. Measurements in the sex glands showed that female phalaropes have unusually large amounts of male sex hormones in their ovaries compared to other birds whose glands where also studied. At the same time two American scientists showed that male sex hormone injected into phalaropes in the dull winter plumage could make them grow new feathers with the bright female colours. Combining our findings it is evident that in late winter, when the birds develop their breeding plumage, females are producing more male sex hormone and are therefore more colourful than the males. The high level of male sex hormone production continues into the breeding season (as my measurements showed) and this accounts for their aggressiveness.

The thought of spring also brings hummingbirds to mind. In Alberta east of the Rockies we are visited only by the ruby-throated hummingbird. Since we have **Ruby-throated Hummingbird** hardly any flowers about our cottage these colourful 3-inch sprites (almost an inch is taken up by the long beak) very rarely visits us. However, two of our neighbours who live in their country cottages all year round and have impressive flower gardens regularly see hummingbirds and they let us know when the hummers put in their first appearance. While females and young males are a glistening green above and white below, mature males have the ruby red throat which gives the species its name. They must breed in our vicinity but so far I have failed to find a nest though I have seen one west of Edmonton on a poplar branch about 15 feet above the ground. It was a little cup hardly one and half inches across made of plant down smoothly bound together with spider webs. After the breeding season a hummer occasionally visits our garden in the city. This may happen as late as the end of September or even early October. In a bird that must be sensitive to cold and still has to migrate at least as far as Mexico for the winter, such late appearances seem surprising.

The drumming of ruffed grouse and snipe are not the only mechanical spring bird sounds. Our woodpeckers including the **Yellow-bellied Sapsucker** flicker also drum. Hammering with their beaks on a suitable branch, they can produce a rolling sound which carries a fair distance. The yellow-bellied sapsucker, a woodpecker which, like the flicker, does not winter with us, drums in a different way from the others. The flicker simply produces a sound that can be represented by "r r r r r". The sapsucker goes "r r r r r -tap - tap -tap".

Starting in mid-April a shrill chorus rings out from every shallow pond and continues for four weeks or more. The main **Chorus Frog** performers in this cacophony are the little chorus frogs only 1 1/2 inches long, with their high-pitched ringing "preep preep" call. The deeper pitched croaking "kowak" or "korak" (similar to the quacking of a duck) of wood frogs is also heard but these are greatly outnumbered by the other species. It seems amazing that these small creatures can keep up their "singing" day and night for several weeks. Only if the temperature falls close to the freezing point, as may happen toward dawn, do they stop for a while. As with all amphibians and reptiles their body temperature follows that of the environment and in the cold their energy production is depressed. Only male frogs sing and strangely enough they do it with closed mouth and nostrils, pushing air back and forth over the vocal ligaments between their throat sacs and the lungs. These throat sacs are pouches connected to the mouth which serve as resonating chambers.

The "song" of the males attracts females which are, at that

Wood frog

season, full of eggs. In a random manner males grasp anything about the right size that comes along. If they seize another male he croaks and is then released, for females remain silent. Apart from her silence a female is recognised by the feel of her egg-distended abdomen. It is reported that males injected with water until their bellies are swollen are seized and held as long as if they were indeed females. (Presumably injected males do not feel like croaking). The male frog grasps the female from behind, passing his forelegs around her chest just behind her front legs. His grasp may help to squeeze out her eggs. It certainly ensures that he is in a position to shed his sperm over the eggs as they emerge. Males stop calling once they have seized a female and the pair remain joined for a considerable amount of time until egg laying is complete. It is not known whether a male after releasing the female starts his song once more, ready to repeat the process with another female. If so his endurance as a "songster" is unmatched by higher animals. There may, however, be a turnover of performers with each one withdrawing after mating.

It is very difficult to catch sight of singing frogs. When one approaches the shore of the pond in which they are performing there is a sudden silence which is maintained as long as one stays close by. Evidently this is a protective response which reduces the likelihood of predators such as herons and bitterns spotting the frogs. An occasional unsatisfactory view of a frog can be obtained from some distance through binoculars. Only its head with the rounded sac under the throat is visible above the water. Better views are said to be obtainable at night with a strong flashlight but I have not been lucky with this method so far.

Our chorus frogs vanish once the spring choral season is over. The only ones I have seen past that season had fallen into the

Wood Frog

vessels I had buried in order to catch shrews. Wood frogs on the other hand are seen all through the warm season and not only near water. They are a mottled yellowish green and have a dark patch behind the eye. This mark distinguishes them from all other frogs. Though they may grow to a length of 3 inches ours are rarely over 2 inches long making them just the right size for a garter snake's mouthful.

Athabasca River

Summer

To the naturalist summer is the time when young birds are already out of the nest, often still being fed by their parents. More exciting to investigate are the northern shorebirds, already on their fall migration, which appear on our lakes and sloughs from July onwards. Migrant wading birds are attracted to muddy and sandy stretches of the shores on inland waters. They show no interest in areas where a dense growth of marsh plants grow right up to the water's edge. In general shorebirds feed on such items as the larvae of midges, mosquitoes and crane flies as well as on small crustaceans. All of these can be found in shallow water. They also look for various kinds of worms and small molluscs, for which they probe the mud. Where marsh plants form the shoreline the water beyond is generally too deep for this method of feeding. Areas that appeal to shore birds are most often found along the edges of shallow lakes in flat terrain like Miquelon Lake. Along the shores of deeper lakes, like Hastings, muddy stretches of shoreline only emerge when the water level is unusually low. When this happens, sandpipers and other shorebirds make their appearance there, too. Miquelon Lake is the best place for shorebirds in our district. I used to imagine it had been named after the island of Miquelon off

Newfoundland which, with its sister island St. Pierre, is the only French possession in North America. These two islands were left to France in 1763 when she lost Canada, to serve as a base for her fishery in that area. However, Miquelon lake was named, according to Holmgren's "2,000 Place Names of Alberta", on a less romantic note after a former post-master of Wetaskiwin.

There are actually three Miquelon Lakes. The largest is the most northern of the three and a provincial park of the same name lies along part of its south shore. This lake is about 4 miles long and 2 miles across at its widest part. All my observations concern this largest of the Miquelon Lakes. Two smaller lakes lie a few miles to the south. Until the 1920's, all three were part of a single large lake. Now only its deepest parts remain as the three isolated waters of the present day.

In an interesting article in the now discontinued magazine "Alberta Lands, Forests, Parks, Wildlife", Edo Nyland tells of the wildlife of this area in earlier days and explains how the lake was reduced to its present state. He reports that up to 1875 buffalo, at times in the thousands, were present in this area as well as plains grizzly and wolves, both carnivores which preyed on the buffalo. Passenger pigeons, which became extinct in the wild about 1900, were then still numerous enough to be trapped for food by some of the early settlers. Within the next ten years, by 1885, all these animals had disappeared. After that period some interesting water birds were still breeding on the lake. There was a large colony of pelicans on one of the islands, but this was deliberately destroyed in the early 1900's by a settler who regarded the pelicans as competitors in his fishing. He landed his pigs on the nesting islands to do the job of destruction for him. Cormorants and blue herons also nested at Miquelon until the 1920's. In 1927 the town of Camrose, 12 miles south of the original single Miquelon Lake, solved a water shortage by digging a canal from the lake into a creek which flows into the town reservoir. This reduced the depth of Miquelon Lake by about 8 feet so that only the present three small lakes remain. Mr. Nyland points out that this water diversion, which had such a harmful effect on the local environment, seems to have been undertaken without permission. The final disappearance of the cormorants from Miquelon Lake was almost certainly due to this lowering of the lake level and its effect on the fish on which these birds depend.

In spite of these adverse changes Miquelon Lake retains its scenic beauty. Pastures run down to its waters. On a sunny day these lie as a glistening blue expanse bounded at the far shore by poplar woods with stands of tall, dark spruce. Beyond these range on range of bluish wooded hills reach the far horizon to meet a brilliant sky. The lake is still a great place for birds. Hundreds of ring-billed and California gulls nest, intermingled, on its largest islands. From here and certain other lakes nearby where they also breed these gulls forage over many miles to refuse dumps as far away as the outskirts of Edmonton. Ring-billed gulls but not the larger California gulls even scavenge at city shopping centers and school grounds. Miquelon Lake also provides breeding habitats for over a hundred pairs of Canada geese. Long-billed marsh wrens and black terns nest in marshes about the lake, and common terns, among our most graceful birds, breed on the small islands.

As in other parts of our district, the mule deer was common around Miquelon until 50 years ago, but the white-tailed deer has

White-tailed Deer

now taken its place. A doe cautiously approaching the lake from the poplar woods followed by her two white-spotted fawns makes a charming picture that may occasionally be seen there early in July. Though they keep well hidden, moose also live in this area and sometimes show themselves.

In 1973 some of the staff of the Miquelon Lake Provincial Park saw what seemed to be three different lynxes including a female with cubs in this area.

Three large shorebirds, the avocet, the marbled godwit and the willet, as well as the smaller killdeer, breed along the margins of

Piping Plover

the lake. It is also one of the few places in Alberta where the piping plover is known to nest. This bird only breeds within a restricted area of central and eastern North America. It reaches the western and northern limits of its breeding distribution in Alberta. A paler version of the larger killdeer, with only a single neckband and a white rump instead of the killdeer's orange one, this rare plover lives and nests on sandy beaches.

If a number of visits to Miquelon Lake are made during the spring and the fall migrations a great variety of sandpipers, plovers and other wading birds can be seen there. The sand brought in to improve the shoreline of the Provincial park provides a beach which is particularly valuable to these birds at a time when little mud is exposed along other parts of the lakeshore. The area about the mouth of a little water-course at one end of the sandy beach generally holds some shorebirds during the migration season, even when visitors are about.

Various kinds of shorebirds are adapted to feeding on different

parts of a lake. Phalaropes hunt water bugs, water beetles and
Phalarope their larvae as well as mosquito and midge larvae while floating over shallow water. They generally keep well out from the shore. While swimming and picking up food phalaropes often spin, turning on their own axis up to fifty times a minute and pecking at the water even more rapidly.

Nearer the shore long-legged dowitchers stand in water up to their bellies probing the mud with their long beaks. These snipe-
Dowitcher like birds have a rich reddish-brown breast during the summer and, at all seasons, a long white patch on the rump by which they can be identified when flying. There are two kinds of dowitchers. The long-billed only passes through our area in the spring and fall for it nests on the western Arctic coast and in Alaska. The short-billed dowitcher breeds in northern Quebec and from the coast of Hudson Bay westwards as far as Alberta. The two dowitchers are virtually in-distinguishable in the field and knowledge of the range of each species was only arrived at on the basis of collected specimens.

Other long-legged shorebirds which feed in shallow water like the dowitcher are the greater and lesser yellowlegs and the stilt
Avocet sandpiper. Large wading birds which nest locally such as the willet, marbled godwit and avocets, also feed in shallow water, often among the others just mentioned. Avocets with their elegant slim shape, their black and white wings and back and cinnamon-coloured head and neck are the showiest of all our waders. They have up-turned beaks which they sweep through the water from side to side obtaining their minute prey by sifting the water rather than by probing the mud like godwits and dowitchers.

At the very edge of the water, but hardly ever entering it, the smaller short-legged sandpipers, often known as peeps, trip up and
Sandpiper down the shore. These are: Bairds', least, semipalmated sandpipers and (paler than all the others) the sanderling. All these, though rarely all of them together, may be seen at Miquelon Lake. Then there is the larger pectoral sandpiper which will not only feed in shallow water and along muddy or sandy shores but also in areas of wet grass such as are favoured by common snipe.

Largely confined to the shore and rarely feeding in water are black-bellied and semipalmated plovers. Both appear in our area
Plover only on migration. They nest further north; the black belly only in the arctic. Golden

plovers may also be seen at Miquelon on their spring passage. However in that season they often rest far from water on pastures or fallow fields.

The golden plovers which pass through the district in the spring to nest in our central and eastern arctic have an interesting pattern of migration. In the fall the old birds move southeast to the Atlantic coast (perhaps because of the abundant crops of berries available there at that season). Then they migrate southwards along the Atlantic coast to South America. On their spring migration, however, they pass northward through the middle of North America along the central flyway. The young of the year seem to have an urge to fly directly southward in the fall so that they move through North America on the central flyway which the adults only use in spring. The locally-breeding killdeer is often to be seen as well on shores which appeal to the other plovers.

Lastly there is the ruddy turnstone, reddish-brown on the back with a complex black and white pattern on its head and neck and orange legs. It nests in the high arctic and appears locally only during its spring and fall passage. It restricts itself to sandy or rocky shores. As its name implies it turns over stones with its beak to get at worms, small molluscs or insects. At Miquelon where there are few stones on the beaches it gets at this sort of prey more often by turning over pieces of stranded water weeds.

Ruddy Turnstone

The most spectacular migration at Miquelon Lake is that of the Northern phalaropes. Unfortunately it only seems to take place under particular conditions which do not happen every year. From 1970 to 1973 there were large flocks, at least 1000 birds altogether, on the lake for several weeks every July and August. They were to be seen floating, each bird about 3 yards from its neighbour, in a flock which covered a large part of the lake. They were easy to tell from Wilson's phalaropes, a few of which were sometimes also about. The northern birds are smaller, have a darker back and a striking black mark which runs through the eye. With few exceptions these birds as well as the Wilson's phalaropes were already in their winter plumage. Occasionally the large flock would be "at sea" off the Provincial Park. I was then able to swim among them. While they were never alarmed by a swimmer they nevertheless kept about six feet between themselves and my bobbing head.

Northern Phalarope

When a flock of phalarope flies up and quickly turns from one side to the other all of them show their white underparts one moment and their dark grey backs the next. They manoeuvre in absolute unison. The same phenomenon is shown by a number of other shore birds. One wonders how the sudden decision to turn is communicated to the whole flock. All its members swing into the new direction at the same time with unfailing precision as if in response to an unseen signal. As present this type of flock behaviour is one of a number of mysteries of bird life.

I believe I discovered why northern phalaropes only frequented the lake during the years I mentioned when I enquired about the level of the lake during the past 15 years from the department of the environment. Water levels were considerably lower i.e. the lake was shallower, from 1969 to 1973 than before or since. Phalaropes, when they feed on a lake, keep about 100 yards off shore. Probably the water there was only shallow enough to harbour their prey during the years of low lake levels.

In recent years it has been found that some female northern and red phalaropes (the latter are very rare inland) behave much like those of the spotted sandpiper. Females having laid one set of eggs will mate with another male and lay another set of eggs. They may even acquire a third mate, all in one season. The male birds, as mentioned earlier, incubate the eggs and look after the young in all three kinds of phalaropes. So far such successive pairings with more than one male have not been shown to take place in Wilson's phalarope, the only one of these birds that nests in our region.

Every summer one can see western grebes, the largest species of the grebe family, on Miquelon Lake. I don't know just where they

Western Grebe nest for these grebes breed in large reed beds such as I have not seen on this lake. They are simply coloured birds, black on top of the head, behind the neck and on the back. Their neck and belly are white. Western grebes have a very long, thin neck. Their high pitched calls (that sound somewhat like "creek creek") carry a long way and often draw attention to the birds when they are far out on the lake.

One more special feature of Miquelon's wildlife should be mentioned; the great grey owl. This largest of our owls and one of the

Great Grey Owl rarest, is a bird of the remote northern forests. There, like the great horned owl, it lays its eggs in the old nests of crows or hawks. In the winter it tends to wander south into the settled parts of the province and is an occasional winter visitor to the Hastings Lake country. In spite of their size these owls do not hunt prey larger than mice or voles. The large, round heads of great grey

Great grey owl

owls with concentric rings on a pale grey facial disk surrounding their yellow eyes make a very impressive sight.

As these owls are birds of the north woods during breeding season I was surprised when I heard that a pair were nesting in a poplar wood at Miquelon Lake in 1972. I was taken to the nest by one of the Provincial Park wardens early in July.

The wood through which we walked to the nest had a number of clearings and these were evidently the favourite hunting areas of the owls. The nest was about 30 feet up in a tall poplar. Two young had hatched from the eggs. The owlets had already left the nest but were still nearby, each perched twenty feet up in the tree. We had a brief view of one of the parent owls but it soon took off and vanished among the trees. The young, more reluctant to take wing, merely looked down at us and snapped their bills from time to time as a threat gesture.

I have not heard any more about great grey owls nesting at Miquelon since that time, but a pair of these birds could easily live unobserved in the extensive woods to the east and south of the lake.

When driving past a dead skunk on the highway one gets a definite and quite sufficient whiff of the characteristic foul smell

Skunk of its anal gland secretion. Apart from our striped skunk, North America has three other kinds of skunks. All of them when seriously en-

Baby skunk

dangered raise their tails and aim a highly offensive spray at the head of the would-be attacker. All have striking black and white colour patterns which are examples of warning colouration. Two distant African relatives of our skunks can also produce a malodorous spray and both these animals also have black and white fur. A predator which has been sprayed once by an animal in one of these unmistakable black and white liveries is unlikely to attack another one. So skunks become immune from disturbance from four-footed predators which have once experienced the stench they can produce. This explains how powerful animals like bears, lynx and cougars have been seen to give up their kill on the approach of a skunk.

Skunks are, therefore, virtually immune from attack by mammalian predators but have no special protection against large birds of prey because these have little or no sense of smell.

Great horned owls are in fact the skunk's most serious enemy. Both the owl and the skunk are largely nocturnal. It may well be that the skunk's wide white stripes make it **Great Horned** easy for the owl to spot it in the poor light **Owl** conditions during which these owls generally hunt. Its warning colouration then works against the skunk in the face of an enemy immune to smells.

One afternoon in mid-April Fred Rourke saw a skunk near his house. It had probably very recently come out of hibernation and did not seem very active. Soon afterwards Fred saw a great horned owl fly into a small poplar directly above the skunk. Here the owl perched for some seconds. With its partly-opened wings hunched forward it uttered a few soft grunts. Suddenly it swooped down and grasped the skunk about the head and neck. The skunk discharged its glands but died without an obvious struggle. As the owl remained on the skunk Fred took a few steps nearer to see things more clearly. The owl clutching the skunk by its head flew up to a height of twelve feet, veered to one side and flew out of sight.

Fortunately skunks don't discharge light-heartedly but only when they feel themselves seriously endangered. An approach at a distance which sends other medium-sized animals away generally does not bother a skunk at all. Even if the disturbance comes closer it does not suddenly run away, as a rabbit might do, but walks away nonchalantly. When it decides to hold its ground it gives warning of readiness to spray by raising its tail and stamping its forefeet on the ground. If the disturber continues to approach, the skunk, still keeping its rear toward the enemy, bends back its head so as to see its enemy and hisses. Unless the opponent defuses the situation at the last moment by standing still (or better yet by withdrawing) the skunk sprays. The spray, always aimed at the face, can reach up to a dozen feet.

On arriving at our cottage one weekend in May we noticed whimpering noises coming from beneath the floor from time to time. It was not long before a definite and unmistakable smell of skunk became noticeable. A quick look around the outside of the house showed two places where some animal had burrowed under the building. Evidently a mother skunk was tending her young under our kitchen. Had it been any other denizen of the wild we would gladly have made this space available as a nursery, but not for skunks. My wife demanded an immediate return to town until the skunks should somehow be expelled. We stopped at our nearest neighbours on the way home, however, to tell them of our unexpected problem. They told us they had heard that mothballs placed in a skunk's quarters would very likely make it go somewhere else. As it happened we had moth balls in the cottage so we went back and flipped the balls through the two holes under the building as far as they would go and resumed our trip back to town.

I did not feel sure that the moth balls would get rid of our skunks so I phoned the pest control officer of our municipality on the following Monday. He promised to set a baited live-trap by our cottage. When we returned on the following weekend the trap was indeed in place but it was empty and the bait was untouched. Inside the house there was no more skunk smell nor any noises coming up through the floor boards. Evidently our unwanted guests disliked the smell of moth balls and had pulled out. I blocked the holes the animal had made with slabs of concrete which prevented this and other skunks from making themselves at home.

A few weeks later we had a much more pleasant encounter with skunks. As we were driving along a secondary road four tiny baby skunks daintily stepped across the pavement in front of us. At this age they don't have the pointed snouts of their parents but are more round headed and look very much like kittens. We did not try to handle them although until they are about six weeks old their anal gland secretion is not offensive. These glands can be removed in a young skunk by a veterinarian and some people keep a skunk so prepared as a pet. But they have little to recommend them as pets. When they get older they are likely to become so fat that they are unsightly and they also become extremely lazy.

Shorebirds are not alone in starting their southward migration in mid-summer. Throughout August and early September flocks of white-fronted geese are likely to be seen passing over, always on a southerly course. They are smaller than Canada geese, brown rather than black and white and have irregular black blotches on their whitish bellies. This gave them the name speckle belly. Their "official" name comes from the white patch on the front of the face and forehead. This patch is not present in birds hatched that year. The white front's call, by which a flock too high for details to be visible can be identified, is less stentorian than that of Canada and snow geese. It is higher pitched, almost always uttered two to three times rather than singly and sounds somewhat like "ka lyo lyok". It somehow suggests laughter.

The flocks we see in the fall come from the tundras of our Arctic mainland and are on their way to the gulf of Mexico where they spend the winter. I have never seen these geese on the ground in the Hastings Lake district, though they rest (in numbers) for a while about Beaverhill Lake. Their spring passage brings them back to that lake but they do not then fly over our district as they do in the fall. I shall always remember hearing the unmistakable call of white-fronted geese long after dark one night in early September. It was clear from the changing sound of the birds that they were not just idly flying about but were coming from the north. They passed almost overhead and forged on southwards.

Some migrants move in complete silence and almost entirely at night. For example the sora rails on our acreage are simply not heard or seen after the last few days of August.

Most of our gulls have white heads all the year round but the two smallest, Franklin's and Bonaparte's gulls, are black-headed **Franklin's Gull** during the breeding season. The story behind their names is almost more interesting than the birds themselves. Franklin's gull has a red beak and legs. In Bonaparte's gull the beak is black and only the legs are red. From late April to mid-September Franklin's gulls are likely to be seen over water throughout our district though they do not nest here. These gulls breed in large colonies in extensive reed beds of large marshes or lakes. Their colony nearest to our area is on Beaverhill Lake. The bird was named in honour of Sir John Franklin who from 1819 onwards conducted the first exploration of much of the coastline of arctic Canda. In 1847 he died somewhere in the area of King William Island. Franklin's gull has the striking word *pipixcan* as its scientific species name. Pipixcan is an Aztec word which suggests Mexico, where this gull was in fact first described. Unfortunately Prof. Wagler, who named the bird about 150 years ago, forgot to record what the word pipixcan means. Unlike the larger white-headed gulls, Franklin's do not come into the city or frequent dumps. But both ring-billed and Franklin's gulls are very quick to notice when a field is being cultivated. They follow the tractor looking for insect larvae and worms in the soil that has been turned over. As soon as work on a particular field is over they leave.

Bonaparte's gull is the only North American gull which usually nests in trees. Its stick nests are built in the tops of spruce in the **Bonaparte's Gull** muskegs of the northern forests. It does not breed in the Hastings Lake area but it appears on our lakes in numbers from late July onwards. From then on Bonaparte's gulls are present, long after Franklin's gulls have departed, and they may be seen well into November when the lakes are already partly frozen. Bonaparte's, unlike Franklin's gulls, do not follow the plough. No doubt they have different food requirements. This gull owes its English name to Prince Charles Lucien Bonaparte, a nephew of Napoleon I. In the 1830's he spent 8 years in North America during which he contributed considerably to the knowledge of bird life on this continent.

Summer in the aspen parkland, despite its attraction for a great variety of birds, brings only one sea duck, the white-winged

White-winged Scoter scoter. This large, black duck has a white patch behind the eye and white on the wing. Like other scoters it spends the winter at sea along both the Atlantic and Pacific coasts. It breeds around most of the larger lakes in the district and stays as late as most of our other ducks in the fall.

I found a nest with five large eggs in a pasture near Hastings Lake one year on July 10. The duck flew up at my approach but could not climb more than 3 feet and after a flight of 15 yards flopped down again. I followed her to get a photo but she flew up once more, only to crash after a further 10 yards. Three days later there were 7 eggs in the nest and it seems likely that this duck was a poor flier because she was carrying an egg almost ready to be laid when I disturbed her.

We have a martin house on a pole near our cottage, in which half a dozen pairs of purple martins rear their young in most years.

Purple Martin Before the martins turn up in May, starlings and house sparrows generally try to take over some of the nesting compartments. This discourages the martins when they come later. We therefore hang a coarse net over the bird house in April and leave it in place until the first martins appear.

One year we had a series of heavy rains in June and early July and during this time we did not see the martins flying about. Soon after the rains had stopped I saw a red squirrel climb the pole, enter one of the holes in the martin house and emerge with a dead young martin in its mouth. I tried to chase the squirrel away with a pole but it dodged from hole to hole and would not leave. No adult martins were in the nest compartments and only one or two occasionally flew near the house. As a result of my harassment of the squirrel, several dead young martins fell to the ground. All were quite cold and there was no trace of blood even on one which had a piece of flesh bitten out of the breast. Evidently the squirrel had not killed these birds but was merely gathering the corpses for food. The young martins had died of starvation during the spell of rain because their parents had not been able to gather enough flying insects. These are virtually their only food.

I had never seen the squirrel near the martin house before. It must have been attracted by the smell of the dead birds. Squirrels

Red Squirrel are not generally thought of as meat eaters, such as the one at our martin house evidently was, but they will in fact eat nearly everything. Mice and voles, young birds, (including occasionally young domestic

American badger
Photo courtesy of National Museum
of Canada

chicks) are part of their diet as well as buds, cones of conifers, fruit and fungi. Fungi are often wedged well up in the fork of a bush or tree by the squirrels so that they dry out rather than decay on the ground. About the time of the martin house episode my neighbour, Joe McKinnon, saw a red squirrel move three young, one at a time, from a bird box where it had had its nest, to new quarters. Joe said that each youngster was carried clinging upside down below the mother's belly, its head between her hind legs and its tail beneath her head.

The largest of our mammals which I have not yet mentioned is the American badger. The Hastings Lake district is near the nor-

American Badger thern limit of its distribution and, as the badger is an animal of open country, it is scarce in the district. On the other hand it is fairly common in the prairie country of southern Alberta. There, when driving along some dirt road, one is quite likely to see one of these animals trotting along the roadside. Its head is marked with a striking pattern of black and white while the body and tail are grey. Stiff hair stands out from the side of the body which therefore looks strangely flattened and very wide, hiding the black legs. The closely-related Old World badger is similar in colour but it lacks this curious flattened appearance. I named our acreage "Badger's croft" in allusion to a character in the book "Wind in the Willows" by Kenneth Grahame. The character was a badger who

127

Summer Scene

was a rather grumpy loner such as I fancied myself to be. At that time I did not know there were any badgers in the district, but over the years I came across their unmistakable burrows from time to time. Though I have yet to see a live one locally, probably due to the fact that they are mainly nocturnal, I heard one moving about in the dead leaves under the trees near our cottage one evening early in October. Next morning a badger was lying dead on the road in front of our place. It had been killed by traffic. It was a male weighing 13 lbs, not a record for they may go up to 25 lbs. Its stomach contained a pocket gopher. This and ground squirrels (gophers), mice and voles are, in fact, the main food of this carnivore. Because of these food habits, the badger is decidedly useful to agriculture.

A mammal far rarer than the badger, which I have never seen in our district but which has occurred here not so long ago, is the

Cougar

cougar. Its normal habitat in Alberta is the Rocky Mountain area and to a lesser extent the southern part of the province including the Cypress Hills. Stragglers still occur in the other two prairie provinces as well. In 1975 there was evidently at least one cougar in our district. At that time a number of local farmers phoned the Alberta Game Farm reporting sightings. They wanted the Game Farm owner to attempt to capture the animal as they feared it might attack their cattle. He already had a family of mountain lions and was not tempted by these propositions. Other cougar sightings were also reported to our local game warden about that time. The daughter of a farmer who lived near Cooking Lake had the good fortune to see a cougar in the open in broad daylight. The description she gave me left no doubt that she had indentified the impressive animal correctly.

In case it might have been an escapee from the Alberta Game Farm, some ten miles away, I enquired whether a cougar had ever got away from there, but I was assured that this had never happened. The only occasion when I have seen a cougar in the wild took place far from the Cooking Lake Uplands. Driving after dark along a secondary road through wooded country in northern California, one of these great cats suddenly appeared in the car lights. It crossed the road, bounded up the steep bank at the side and was gone in seconds.

There are woodchucks in our district as in the rest of Alberta except the mountains and the southern prairies. Judging from my **Woodchuck** own observations this large brown rodent is rare, for several years may pass before I see even one. One spring a woodchuck made its temporary home under our cottage and through the kitchen window we could see it come out to nibble the grass by the woodpile. His was a short stay for we never saw him again.

We are fortunate in having no Norway rats or house rats. These loathsome and ugly rodents reached our continent in ships from **Norway Rat** the Old World in the late seventeen hundreds. It is only during the present century that they have spread westward across the prairies. A constant, vigorous control program in settlements and farms along the eastern border of Alberta has so far prevented them from invading the rest of the province. These rats not only consume stored grain but also spoil much of what they don't eat with their excreta. I shall always remember the unpleasant sight of these animals, with their ugly long bare tails, scurrying about under the benches in St. James' Park in London. They were working over the scraps left by visitors. No mammal pest is more harmful to man than the Norway rat. Not only do they devour foodstuffs worth about $20 million a year in Canada but they sometimes carry fleas infected with the plague bacillus which can be conveyed to a person by a bite from such a flea. They also act as reservoirs of other dangerous microorganisms.

Another, but less harmful, rodent inadvertently introduced into the Americas by man is the house mouse. It is generally distributed over the settled areas of parkland but is **House Mouse** nowhere really common. I have only once caught one in the two traps generally set out in our cottage. With its uniform brownish-grey colour and long sparsely-haired tail it is easily distinguished from our native mice.

To find out just what small mammals lived on our acreage I set out a series of mouse traps for several months because most of these little animals can only be iden- **Deer Mouse** tified in the hand. In the case of shrews it is even necessary to examine the teeth with a magnifying lens. My outdoor trapping yielded deer mice, red-backed voles and meadow voles in approximately equal numbers. However, deer mice were the only kind that often entered the cottage. As they can climb amazingly well and leave their droppings wherever they go we do not welcome them indoors and always set traps for them in the kitchen. Though not welcome indoors we never-theless admire their good looks. They have large, rounded ears and big eyes, are greyish-brown above and white below with a decently haired tail neatly divided along its length into a grey upper and white lower half.

Shrews are, if anything, more secretive than mice and voles. If one wants to determine which kinds occur in a particular area one must capture them. I had no luck with this **Shrew** as long as I used mouse traps, though I baited them with bacon which these little meat-eating insectivores are supposed to favour. Acting on a tip from the technician in charge of the Zoology Museum at the University I buried tall glass jars, the kind in which instant coffee is sold, so that the open end was flush with the ground. Strategic placing of such jars soon yielded a sample of my shrew population. Masked shrews were the most common. Next in numbers were Arctic shrews. I also caught a few, very few, of the rare pigmy shrews, the smallest North American mammal. It is only 3 inches long from the nose to the end of the tail and weighs, on average just under 3 grams or 1/9 of an ounce.

To return to rodents; there is the pocket gopher, a compact, blunt-headed brown burrowing creature 8 inches long with its short tail included. The heaps of earth **Pocket Gopher** thrown up by pocket gophers in the course of their excavation of underground passages, inches to a foot below the surface, are common throughout the district. But the animal itself is rarely seen for it pushes up soil from its burrow at

*Richardson's ground
squirrel (gopher)*

night. In their underground passages pocket gophers feed on the roots of a variety of plants but at night they come above ground to gather green plants which they cram into their cheek pouches and eat at leisure underground. In the fall they take a store of roots into their burrows, for they remain active through the winter. These little animals expose themselves above ground so rarely during the day that I have yet to see one, though my wife once came across one busily digging. I have also had no luck catching them in mouse traps, though I understand they can be taken in special traps set in their burrows. A neighbour's cat often brings a dead pocket gopher to her owners and in this way I get all the specimens I might want. They show the pouches which open on the face beside the mouth and extend back some distance. The size of the pouches seemed to justify the rodent's name.

Franklin's ground squirrel is a characteristic dweller of the Aspen Parklands. It is not found in the coniferous forest of the north and west nor on the prairies.
Franklin's Ground Squirrel Somewhat larger than our common gopher (Richardson's ground squirrel) it is also more grey in colour and has a bushy tail. This omniverous ground squirrel was named in honour of Sir John Franklin who, after commanding two successful expeditions to the Canadian Arctic, perished with his crew on the third expedition in 1846. Richardson's ground squirrel gets its name from Sir John Richard-

131

son, surgeon-naturalist on one of Franklin's explorations and on one of the search expeditions undertaken after that explorer's disappearance.

While gophers live in the open the Franklin squirrel seldom wanders far from trees or shrubs. Apparently it spends most of its time, even in summer, below ground in one of its burrows. Because of this and because of the cover in which it moves when on the surface, it is seldom seen. The eating habits of this 13-inch long (including tail) animal are not altogether innocent from our point of view. The vegetable part of its food includes cultivated grains. But it is also carnivorous, taking insects, caterpillars, frogs and toads. All would be well if that completed the list of its meat diet but it also takes the eggs and small young of ground nesting birds. In the Delta Marsh in Manitoba these squirrels robbed almost as many duck nests of their eggs as did crows. It is also believed they do more damage in gardens than any of our other "gophers". It is just as well that they are fairly rare in our area.

The uplands are well within the range of the least chipmunk but I must confess that I have yet to see one there. It has been recorded in Elk Island National Park but was listed as uncommon.

In early August, bank swallows gather in large flocks near some of our lakes. One sunny day early in that month we saw hundreds

Bank Swallow
of them, with a smaller number of tree swallows, resting between short flights on the telephone wires along the secondary road which runs along the shore of Hastings Lake. Many were crouched on the paved road, evidently sunning themselves. Further along the lake we came across two more large flocks of bank swallows resting on the wires. At other times we have seen barn swallows among the others in these summer flocks.

There are more bank swallows in these flocks than nest in the district, so they must represent a gathering of birds from a large area. They are probably collecting as a prelude to their southward migration. Bank and tree swallows leave about the end of August but barn swallows stay well into September.

In August there is also a movement through our area of certain songbirds from the north. They are seen singly or, at the most, in

Least Flycatcher
twos or threes and generally keep to the cover of trees and bushes. Among birds in this category we have on occasion seen the colourful western tanager, more often the yellow-rumped warbler, as well as the less conspicuous grey-checked thrush and the olive-sided flycatcher.

Our own flycatcher, which nests freely on poplars, is the smaller least flycatcher.

In July and August there is also some wandering about of birds which do not migrate. In this category are the red-breasted and white-breasted nuthatches of which we see **Kingfisher** one or two during this season in some but not all years. A wanderer which is also a migrant is the kingfisher. One made itself at home by our dugout for a day in July of two different years. From a convenient perch it dived repeatedly into the water. Kingfishers like small fish but as there are none in our dugout it must have been satisfied with tadpoles, salamander larvae, water beetles and perhaps water snails.

Looking over some lake-side cottages which had been empty for some time I have more than once come across the dried and shrivelled body of a female goldeneye duck **Goldeneye** near the open fireplace. Goldeneyes, as well **Duck** as buffle-heads, nest in holes in trees. The holes may be natural or made by woodpeckers. In both cases the downy young scramble out of the hole soon after hatching and make the drop of sometimes twenty feet to the ground unharmed. To some hen goldeneyes the horizontal hole of a chimney evidently appears as good a place to nest as a tree hole. They descend the chimney. From the bottom they can neither fly nor climb out again and so die of starvation.

On a sunny day in July a half-grown ruffed grouse flew into a window of our cottage with such force that it broke the glass and was killed. It is not surprising that fast fly-**Ruffed Grouse** ing birds occasionally collide with a window when they perceive an apparent passage behind it. However, ruffed grouse not only fly into windows but also into the walls of houses. I have not heard of such behaviour from other birds. Sometimes the grouse only stun themselves but occasionally the impact kills them. Two Saskatchewan farmers have kept records of this unusual behaviour. They found that it could happen at any time of the year. Their explanation is that when grouse flush and fly off, usually in a straight line low over the ground at about 50 m.p.h., they can avoid natural obstacles like tree trunks by a slight swerve to one side or the other. Such a swerve will not however save them from an impact with a long object like a house unless the grouse perceives it from a good way off.

There are still a number of summer birds that I haven't mentioned. I will restrict this account to just a few. The broad-winged

Broadwinged Hawk hawk has much the same build as its larger relative, the red-tailed hawk. It favours woodlands, whether they be pure poplar stands or mixed woods, and avoids open country. With its liking for woods, the broad-wing is quite characteristic of the Hastings Lake district, yet until this year when a pair built on our acreage I had never seen its nest. Because it builds anew each year, the nests do not become bulky and easy to spot in winter like those of the red-tail.

Broad-wings are far more secretive than their larger relatives. The nest I found was over forty feet high in the fork of a tall poplar. Only one feature distinguished it from the Cooper's Hawk nest I had seen. That was a poplar twig with green leaves drooped over the side of the nest. The habit of decorating their nests is characteristic of the buteo hawks, the group represented in our area by the red-tailed, Swainson's and broad-winged hawks. When I first spotted the nest with binoculars I could just make out the top of the head and the yellow eyes of the sitting bird. We looked at one another for a few moments. Then I walked away quietly, not wanting to scare it. When there are only eggs in the nest hawks readily desert it on disturbance, usually to start a new nest. If they are disturbed after the young have hatched they are more prepared to carry on in spite of interference.

I looked at the broad-wing's nest a few more times from a reasonable distance and was always able to make out some part of the sitting bird. Yet a week later the nest was emply. No hawk flushed from it even when I tapped the nest tree with a stout stick. The pair had evidently deserted. Yet I still caught an occasional glimpse of one of the hawks and heard the characteristic call, so high-pitched it could be called a whistling "tsee hoo ee". This suggested that a new nest might not be far off. Staring up at every tall poplar on our place I did indeed find a heap of sticks I had not seen before high up in one of our trees. It was a nest, evidently still unfinished, but it had a small tell-tale bunch of leaves. However, up to the time of writing, I have not yet seen a hawk on or near this nest.

Unlike red-tails, broad-wings do not perch in treetops but somewhat lower down and they rarely soar, even during the breeding season. They can hunt by flying through the woods hoping to come on prey unaware. More often they hunt by watching the ground from a high perch waiting for a mouse or vole to appear. These hawks can be very patient. I watched one hunting in the manner just described for a long time during which nothing showed up to make him pounce. Only at the end of an hour did he give up and fly elsewhere. Of course they are sometimes luckier than that particular bird. Just the other day as I was coming up our driveway a broad-winged hawk rose from the ground and flew away with a good sized garter snake in its talons. A meal like that would probably keep it satisfied for three days at least. As this hawk rose up in front of the car with its back to me, I just had time to take in the light and dark bands across its tail and the particularly wide dark band at the end, which is the hall-mark of this species.

Broadwinged hawks arrive in the spring nearly a month later than the red-tails. They also leave us earlier in the fall. Their migrations are extensive and take them to central America and into South America as far as Brazil and Peru.

In the fall many hawks on their southern migration tend to fly along the eastern edge of the Appalachian mountains. Considerable numbers of them pass certain favoured look-outs. Twenty years ago I was able to see part of the hawk migration from one of these points, Mount Tom in Massachusetts. These birds pass over Mount Tom in much more modest numbers than they do over some of the more famous "hawk mountains". Even so, during five hours of watching on September 16 I counted 505 broad-winged hawks, 7 ospreys, 2 marsh hawks, a Cooper's hawk and a merlin. The broad-wings were circling in a loose mass roughly forming a cylinder. They were evidently riding an up-draft. Gradually individuals or small groups would glide away from the mass of soaring birds and sail to the southwest, keeping parallel to the mountains on their right. On another day at Mount Tom I saw, apart from broad-wings, a sharp-shinned hawk, a red-shouldered, 4 red-tailed hawks and 6 sparrowhawks or kestrels.

The migration of hawks calls to mind an experience I had in Argentina. It concerns Swainson's hawk which breeds in our area,

Swainson's Hawk though, being a bird of the open plains, does not live in the Hastings Lake district. It is kindred of the red-tailed hawk but winters further south going as far as the plains of Argentina. In 1973 I had seen Swainson's hawks at home in early September, their usual time of departure from central Alberta. Soon after that I was on the pampas of Argentina doing field work every day. It was not until the 29 of November that I saw a single Swainson's hawk. When early in December I saw a flock of them overhead it seemed like a greeting from home. It had taken them two months

Black bear

to fly over half of North America, over Central and half of South America.

This attractive hawk was named after William Swainson (1789-1855) who, despite being a first-rate naturalist and fairly good artist, had a rather chequered career. Born and raised in England, his father got him a job in the colonial service. He retired from this after eight years to become a fulltime naturalist. He published several books on zoology, among them the well-known *Fauna Boreali Americana* (an illustrated account of the birds and mammals of northern and arctic North America) of which he was co-author with Sir John Richardson. Yet he was unable to support his family and finally emigrated to New Zealand in 1837. There he became a farmer, taught natural history and did some more writing. He remained in New Zealand until his death.

In spring or summer, though extremely rarely, a black bear wanders into the uplands. I was told of one which raided the garbage beside a local RCMP constable's house

Black Bear one night in 1973 or 1974. Al Oeming has told me that a black bear wandered into his game farm (now Polar Park) on several occasions and another informant reported one seen near the Blackfoot forest reserve. Black bears are found regularly in the densely forested country some 130 miles to the north and west and it is from there that those seen in our area must have wandered.

Personally I feel it is just as well that black bears are so rare in my district because they can be dangerous. It is true that in the vast majority of instances a truly wild black bear takes off when it becomes aware of a person. (Those near the National Park townsites are more used to humans and have largely learnt to suppress their escape reaction at the sight of people.) However, attacks on people have occured. In August, 1980, a black bear killed a man near Zama in the Hay Lake area of Northern Alberta. Shortly after this the same bear attacked a couple walking in the woods. The two climbed a tree but the bear pulled the woman down and killed her. Eventually this bear was shot and found to be quite normal. Its behaviour could not be attributed to disease or some injury which might have prevented it from feeding normally.

Within days of the Zama episode a black bear tore through the wall of a tent occupied by several people and severely mauled an old woman. This took place in the Lesser Slave Lake area. There are also at least two records of children killed by black bears. While these are all very rare instances the fact remains that bears, whether blacks or grizzlies, are unpredictable.

A young friend and I had an experience which looked dangerous for some moments but ended on a humorous note. We were walking, unarmed, along the shore of a lake in the very northeast corner of Alberta. A black bear, not quite fully grown, suddenly crashed out of some bushes and took off at a run. When about sixty yards from us he turned around and to our dismay came toward us in a lolloping gallop. My companion grabbed a long stick and waved it about while I yelled at the top of my voice. The bear stopped twenty yards from us, turned around and ran away. It had probably been motivated by mere curiosity, but might not a fully grown bear have continued to come at us in spite of our noise and gestures? He too could have been merely curious and might have stopped on catching our scent, or he might not.

Whenever I walk alone along some trail in the boreal forest I am aware of a slight, a very slight, uneasiness which compels me to turn around and check what is behind me from time to time. The only rational basis for this feeling is the remote possibility that a black bear might show up and the even more remote possibility that it might cause trouble. However I think this almost subconscious apprehension is mainly irrational. It is probably an inborn reaction, inherited from our distant ancestors, to an environment in which all kinds of dangers lurked behind a screen of trees and undergrowth.

Not all black bears are black. Brown ones are fairly common

along the Canadian segment of the rockies. Grizzlies also occur in
that area. They resemble the so-called cin-
Grizzly Bear namon bears (brown-black bears) to some
degree. Grizzlies have a pronounced shoulder hump lacking in the
other species and a dish faced profile. The profile of the black bear
is straight or shows a Roman nose. Tracks are very helpful to indicate the presence of grizzlies. The long claws on the grizzly's
forefoot leave imprints fully an inch ahead of the toe prints
whereas these prints are less than half an inch apart in the black
bear track.

Another colour phase of the black bear is found on Gribble
Island in British Columbia, where many of the bears are almost
white. There are also so-called blue bears (actually lead grey in colour) of Southern Alaska and adjacent parts of the Yukon. Here, as
is often the case in the names of birds and mammals, colour
designations are exaggerated. Greys are called blues, and reddish
or rusty browns are called red.

Two or more colour phases occur in several other species of
animals. The black and the cross phase of the red fox are well-
known examples. Among birds the blue goose, formerly thought
to be a distinct species and now known to be merely a colour
phase of the lesser snow goose, is a striking instance. Animals of
different colour phases can and often do interbreed. They are after
all merely differently coloured individuals of the same species.
There are always a number of animals of a particular phase. Their
colour is thus not simply due to some individual abnormality of
colour development but is the expression of a genetic feature
shared by many individuals.

Bears are classed among the carnivores, yet three quarters of the
black bear's diet consists of plant material. In spring it will take
fresh grass, certain roots and even spruce needles. Later in the
season it finds berries and hazelnuts. Soon after the bears have
emerged from their winter sleep they often find carcasses of
winter-killed big game animals. Carrion actually forms the bulk of
their fresh food although they will also eat small animals like mice,
snowshoe hares or woodchucks. Insects and other invertebrates
contribute to their protein requirements as well. Occasionally they
kill young moose, elk or deer. Bears that live near settled areas
sometimes attack calves or piglets and raid beehives.

Black bears in our region are dormant from late October to early April. They spend this period in a winter den which may be
under logs, in a cave, a large hollow log or in some other sheltered
site. These shelters become covered with snow except for a hole

which is kept open by the bear's warm breath. In rodent hibernators like woodchucks and ground squirrels the body temperature drops from the normal 36°C to 3-5°C while they are in the dormant state. In black bears it is only reduced to 31-34°C and the animal can be aroused by a loud noise. Body functions in a dormant bear are evidently reduced to a lesser degree than they are in hibernating rodents and it is debatable whether the black bears' winter condition should be considered true hibernation.

Female black bears are only in heat for about three weeks, from late June to early July. Only during this period do the bears live in pairs. Males lead solitary lives for the rest of the year. Females on the other hand are accompanied by young for a good part of their lives. The young of nearly all our large mammals are born in spring or early summer when the climate is most favourable and food is most abundant. Yet bear sows (grizzly and polar as well as black bears) give birth in midwinter while they are in their dens. The newborn cubs are minute, those of black bears weighing only about 10 ozs. (300 g), one six hundredth of their mother's weight of perhaps 440 lbs. (200 kg). Perhaps this explains their winter birth. The extremely small cubs are undoubtedly very sensitive to cold. They are probably in a warmer environment beside their mother in her snug den than if they were born in summer when the mother, then in an active state, would leave them exposed whenever she went in search of food. But this is mere speculation.

Writing about ten years ago A.W.F. Banfield in his "Mammals of Canada" wrote that the average price of a black bear pelt was $24.74. This hardly seems worth the effort involved in skinning and preparing it. He states that these days there is little use for the pelts except for the fur bushes of certain Guards units (in Canada, Britain and Denmark) and of members of pipe bands. No doubt because of this connection there is a lively pipe and drum march called "The Black Bear". A distinction for this bear not matched by the otherwise superior grizzly.

Autumn

For some weeks in the fall, duck hunting, which I pursue ardently though with only modest success, takes up some of the time I
Muskrat would otherwise spend observing. But my early morning and evening hours at the Hastings Lake narrows are often rewarded by interesting sights other than ducks. As I stand on the shore with the background of a dense willow thicket, muskrat and beaver swim past unconcerned. Only if I move will a beaver slap his tail in alarm and dive. It was at the narrows that I discovered that muskrats have quite a pleasant soft call which they sometimes uttered while gathering vegetation near the shore. I have found only one book which mentions that muskrats have a voice. The author describes it as a squeeking which he had only rarely heard.

Occasionally a goshawk comes flying over quite low as he crosses the narrows from the north, or an early bald eagle flaps and glides over the trees of the opposite shore for a while.

One evening when I waited for ducks on an open part of the lake shore, a buck and five does came out onto the pasture behind me. They were white-tailed deer and fascinatingly graceful.

On a September day, we saw a peregrine falcon winging its way southward. This migrant has become very rare over most of North
Peregrine America due to the use of pesticides, par-
Falcon ticularly DDT. The poison becomes con-
centrated in birds the falcons eat and causes them to lay abnormally soft-shelled eggs which fail to hatch. Before pesticides became widely used peregrine falcons regularly nested in Alberta. In those far off days, I recall seeing a nest with eggs on a steep mudbank above Whitemud Creek, well inside the present city limits. Nowadays there are only a few pairs of these noble falcons breeding in Alberta. Those we see passing through in the spring and fall nest in the Arctic where pesticides have not yet been used and where there is still a normal population of these falcons. In Britain peregrines became almost extinct some 30 years ago. Prohibition of the use of particularly harmful pesticides brought about a remarkable recovery of the population. The simple control of pesticides was successful in this case because in Britain the falcons hardly migrate, staying in the country all the year.

In Canada the situation is more complicated. Falcons from middle and southern Canada may winter in Central and South America where DDT is still used. The control of pesticides within

Canada and the U.S.A. therefore cannot protect the birds all year round. To help the falcons the Canadian Wildlife Service is involved in a vigorous programme of raising them in captivity at Wainwright, Alberta. Full-fledged young are liberated from time to time, some of them on city high-rise buildings. To the falcons these are cliffs, from which they have a chance to catch domestic pigeons and where they are safe from predators.

It was a pleasant interlude when on a recent visit to my dentist's office in downtown Edmonton we saw one of these falcons dash high across the street to land on a sign which formed a convenient perch on a nearby high-rise building. The C.W.S. programme can build up our peregrine population with the hope that a number of the liberated falcons will bring off a few broods of young before the pesticides they absorb while wintering in the south make them sterile. Whatever happens, as long as the Canadian and Alaskan Arctic remain in their natural state we will not lose this dashing hunter of the skies.

I mentioned that a pair of Cooper's Hawks have nested in a woodlot near our acreage for several years. These hawks will not tolerate the smaller sharp-shinned hawk nesting within a 1 1/2 mile radius of their own nest. This is because sharp-shins hunt in the same manner as Cooper's Hawks and the two species therefore compete for much the same prey. As a result we only see sharp-shinned hawks passing over singly in the spring and fall. When doing so, they generally circle at some height, gradually drifting north or southwards according to the season. Sharp-shin, Cooper's hawk and the goshawk are all closely related. All have long tails and short rounded wings. They all hunt by flying at a low level from cover to cover and taking their prey by surprise.

Sharp-shinned Hawk

There are no English names for the various groups of closely related hawks. Bird watchers have therefore been compelled to use the scientific generic or group names for these birds. I have already referred to the name *buteo* for the broad-winged hawk and its relatives. The goshawk, Cooper's hawk and the sharp-shin belong to the group called *accipiters*. The names buteo and accipiter come from the book on natural history published in 77 A.D. by the Roman Gaius Plinius. His book contains much false as well as true information so that we cannot really tell what sort of bird of prey his buteo and accipiter represented but later these names have been attached to well-known species. Plinius, though an unreliable naturalist by modern standards, was a man of many parts. In his younger days while commander of a cavalry regiment he wrote a treatise on missiles to be thrown from horseback and a history of the wars between Rome and the German tribes. Later he produced a work on oratory and another on grammar. For some years he was procurator of Spain. He ended as commander of a fleet off the Italian coast at a time when Mount Vesuvius was in a great eruption. Here his curiosity about natural phenomena, which had inspired his "Natural History", led to his doom. He landed too near to the volcano and succumbed to its stifling fumes.

There are two birds which are with us all the year round which should be mentioned here. The first is that familiar bird, the house or English sparrow. It is a distant relative of

English Sparrow the tropical weaver birds and is thus no kin to our own native sparrows. Introduced from England in New York in 1850 by some mistaken sentimentalist, these sparrows have since spread all over North America south of the Arctic. They manage to survive the hard winters in places as far north as Fort Simpson on the Mackenzie, and Churchill on Hudson Bay. I came across house sparrows in Argentina and in Australia. In fact the only large settled areas they have not managed to take over are China and Japan. There they are replaced by the rather more attractive Old World tree sparrow. Originally, before various introductions, the house sparrow was confined to Europe, western Asia and North Africa. I can hardly think of anything good to say about house sparrows — they are messy, they crowd out other more attractive birds, and they have no sweet song. In winter, however, when our bird life is reduced to near zero, their antics and their chirpings do enliven the scene. They desert our acreage, where some always nest, in mid-winter but on sunny days when it seems their sex drive re-awakens, they come back for a day or so. I thought that I should try to get really interested in house sparrows. After all, along with magpies they are almost the only birds one can be sure to see here in winter. So I bought a book by a British biologist which was devoted entirely to this common bird. It turned out to be the first bird book I never finished.

The starling is not quite as bad as the house sparrow. It too was introduced to this continent from its original home in Europe and parts of Asia, in 1890. It spread more slow-

Starling ly than the sparrow over much of the U.S. and Canada reaching southern Alberta about 1934, but it did not reach the Edmonton area until 1948. Its range does not extend as far northward as the house sparrow's. In our region it reaches its limit at Ft. Smith. As in the case of the sparrow, some people in other parts of the world have also felt an urge to add the starling to their native bird life. It is now also found in the West Indies, South Africa and New Zealand.

Starlings in their glossy black breeding plumage with its green or purple reflections are quite handsome, though the way they waddle about on their short legs is hardly graceful. Their songs are a sequence of whistles, chirps and squeeks, often including very skillful imitations of a variety of other birds. They will sing on sunny days even in winter and it is then that their not-very-musical but cheerful notes are most likely to be appreciated.

Starlings can mimic the red-winged blackbird's song so faithfully that when I hear the first redwing's song of the spring signaling the arrival of these birds, I make sure that I was not in fact listening to a starling. Starlings can also imitate the screech of the red-tailed hawk and the unique "whee weeoo" call of the drake baldpate in the most convincing manner.

Young out of the nest are dull if not ugly looking. They are greyish brown with a whitish throat. Their continual harsh "tsharr" calls are decidedly unpleasant and where these ugly youngsters are numerous the chorus of their rasping calls can get on one's nerves.

Nearly all starlings leave our area for the winter. There are certainly none on our acreage during the coldest months. Small numbers of starlings do however manage to survive the winter in our northern cities. A habit that may help them to stand severe cold is perching on chimney tops with their backs to the updraft of warm air. The majority that went south for the winter come back in March well before our other small migrants show up. This gives them an edge over our other hole-nesting birds such as bluebirds, tree swallows and purple martins. Starlings nest only in holes and due to their early arrival they generally manage to get the best ones.

One point in favour of this common invader is its diet. Among the insects it eats there are many that are harmful to agriculture.

I had known for some years that there were mink in our district. My friend Fred Rourke whose house overlooks Hastings Lake, had

Mink

told me that he occasionally saw one on the lake shore or, when boating, he sometimes came across a mink swimming in the lake. Like so many of our mammals, mink are mostly active at night and during twilight. This explains why for a long time I occasionally came across their tracks but never saw one of the animals. One evening early in September as I was returning to our acreage on horseback, I saw a dark coloured hunched-back animal loping along the roadside. I

143

put the horse to a trot to get a closer view and the animal came out onto the middle of the road and stayed there until we were only about fifteen yards apart. It was unmistakably a mink. The characteristic white spot under its chin was plainly visible. For some seconds it seemed uncertain which way to go but finally moved to the west and vanished among the roadside weeds. It should have stayed there, for it had killed some of my neighbour's chickens and he was on the lookout for its return. It did return and was shot later that very evening.

The mink proved to be a large male 580 cms. long compared to an average length of 540 cms. Females are always somewhat smaller. Normally birds form a very small part of the mink's diet. It lives mainly on small rodents including muskrats, fish and frogs.

Among the birds which pass through our district to and from more northern breeding quarters several are most prominent in their fall passage. This is certainly true of

Bald Eagle the bald eagle. The first of these impressive birds may sail onto a tree top along the shores of Hastings Lake as early as September. It will spend much time watching from its perch or taking short low-level flights over the water and will probably leave after a few days. The real influx of bald eagles at Hastings Lake is not until mid-October. Thereafter there will be some of these great birds, sometimes as many as a dozen, until the lake freezes over completely in late November or early December. Eagles that are seen coming in from a distance and not merely flying about over the lake almost always fly in from the northeast. This suggests that they come from northern Saskatchewan which has a greater breeding population of these birds than any part of Alberta.

Eagles can only find food on the lake as long as there is some open water. Once the freeze-up is complete they take off, flying steadily to the west or southwest, until they are lost to sight. Some of them do not have to go very far to find suitable winter quarters. The stretch of water kept open by the outflow from the power plant on Lake Wabamun (west of Edmonton) generally supports two or three eagles through the winter. Others winter along the Bow river near Calgary. Most of them, however, travel further to the Pacific coast or to inland waters in the U.S.A.

An adult bald eagle, black save for its white-feathered (not bald) head and white tail, gliding or slowly flapping its huge wings against a blue sky is a magnificent sight.

Golden Eagle This two-tone plumage is not fully developed until the birds are four years old. Young bald eagles are simply brownish-black all over and, unlike the adults, they are not easy to distinguish from the all-brown golden eagle. Single golden eagles also pass through our district in spring and fall but they are much rarer. Unlike bald eagles they are not particularly attracted to water nor do they stay any length of time. In good light an adult golden eagle shows tawny or golden brown on the back of the neck. This accounts for its name, for the rest of the bird is simply dark brown. Immature golden eagles can be told from the all-dark young bald eagles by a patch of white on the wing and some white at the base of the tail.

Bald and golden eagles are not close relatives. The bald eagle is our sole representative of the sea eagles which include the somewhat larger white-tailed sea eagle of Europe and other parts of the Old World and the huge Steller's sea eagle of northeast Asia, 1 1/2 times as large as the bald eagle.

The bald eagles that are to be seen for six weeks every fall at Hastings Lake may hope to find a fish washed up on the ice edge of a water hole but there are so few fish in the lake the chances of this happening are very small. Mainly they count on catching ducks at the water holes; ducks that were crippled during the hunting season or occasionally uninjured ones. A duck in full flight can usually out-distance the slow flapping eagle. In order to be successful the eagles need infinite patience coupled with a readiness for instant action when a favourable opportunity arises. The nearest I have come to seeing one of the Hastings Lake bald eagles secure its prey was on a late November morning when I was watching a flock of about fifty Canada geese resting on the ice. Suddenly they flew up and when I switched my eyes back to where they had been, an adult bald eagle I had not seen before was on the ice pecking at a small dark object which may have been a duck or a muskrat. The eagle coming in from a distance had flushed the geese but it was not one of them it was about to surprise. It was interesting to note that no sooner had the first eagle started tearing at its food than another adult sailed in and landed alongside the first.

It is not uncommon to see one or two eagles perched on the edge of the ice around a stretch of open water while ducks swim quite unconcerned within a few feet of the hulking predators. Yet these ducks are not as safe as they seem to be. My friend Dick Dekker, whose patience in observing the eagles equals their own, has seen several of their successful hunts and described them to me. On one occasion some ducks were resting on the ice a few yards from the water's edge, perhaps because the ice margin was too thin to bear

145

their weight. From time to time an eagle would fly low over them. At every pass the eagle made the ducks scrambled into the water where they could avoid its attacks by diving. But one scaup duck was too slow. The eagle put down a foot and simply hooked it up.

On another occasion two mallards flew up at an eagle's approach. One of the ducks, instead of flying away, tried to fly steeply upwards. The eagle, with powerful wing beats, climbed above it then suddenly dropped onto the duck and seized it.

At a small water hole Dick saw an eagle hovering over a mallard which dived whenever the eagle came too close. Finally the duck became exhausted and was grabbed by the predator. At this point another eagle attacked the first which dropped the duck into the water and moved on. The second eagle forced the duck to dive to exhaustion again and finally seized it for good.

But such lively goings-on are rare. I shall always remember the doleful scene of five all-black, immature bald eagles spaced along the edge of a small hole in the ice while snow flakes were drifting down. They were at the only opening in the ice on their part of the lake and there was not a single duck in the area. Their chances of getting something to eat at that spot were zero. In the snow falling from a grey sky they presented the very picture of sombre desolation. Another immature bald eagle I saw late last fall showed more sense. It first appeared some distance from the lake flying in from the east. Nearly an hour later when I arrived at the lake, there it was standing on the ice at a spot where one could see that there had earlier been an open hole. The lake was by then completely frozen and after another 45 minutes the eagle took wing and flew off steadily to the west till it was lost to sight. It was the last bald eagle of that year.

A number of years earlier when Hastings Lake was still open in November I saw an immature bald eagle perform a feat I would have thought impossible. I was waiting for passing ducks when the eagle left its tree perch on the opposite shore and coming toward me settled on the water with outstretched wings. There was a dark object in the water, possibly a dead duck, but the eagle did not touch it. After a short while it flew up without any difficulty and returned to its tree. Soon it was back again and landed on the water at the same spot as before. This time it kept its wings closed and for several minutes it floated high in the water much like a gull until it flew up and left for good. I thought I'd seen something unique but in Bent's "Life History of North American Birds of Prey" I found the statement "bald eagles have been seen to alight on water on several occasions, floating for some minutes lightly as a gull, probably in pursuit of fish (this did not apply in my observation) then arising from the surface with no great difficulty".

Adult bald eagles can look magnificent but their choice as the national emblem of the U.S. was unfortunate. As Bent, himself an American, records, this eagle's carrion feeding, its cowardly behaviour and its attacks on the smaller and weaker osprey (which it forces to drop its prey) do not exemplify the best in the American character.

The golden eagle has long been considered a symbol of nobility and undaunted fighting spirit. It will tackle much more powerful prey, such as roe deer and chamois, than a bald eagle would dare attack. When its talons are struck into a large animal it can keep its grip while allowing itself to be dragged along for as much as a hundred yards. The trained golden eagles of Central Asiatic horsemen will even kill wolves.

The supreme god of the ancients was represented in the classical world with an eagle at the foot of his throne. A silver eagle was the standard of every Roman legion and, as in the American bald eagle emblem, it carried a bundle of arrows representing bolts of lightening in its talons. The Roman eagles were symbols of valor and national honor. When a Roman army was defeated in Syria and a number of eagle standards captured by the enemy, this loss was felt as acutely as the defeat. When, years later, the "eagles" were returned by the king of the former victors, the event was celebrated in verse by the poet Ovid. The eagle of Imperial Rome made a lasting impression on many people and it became the eagle of heraldry. The Byzantine emperors of Constantinople, successors of the Roman empire, used a double-headed eagle as their seal. The two heads symbolised their claim to rule over the western and eastern worlds. When the Turks overran Byzantium, the rulers of Moscow considered themselves its heir, and they and their successors, the tsars of Russia, adopted the double-headed eagle emblem. Further west Charlemange, who founded the Holy Roman empire as a successor to the western Roman empire, adopted a double-headed eagle on a yellow background. In due course this became the emblem of the emperors of Austria. A single-headed black eagle was the emblem of Prussia and it survives in the coat of arms of the West German republic. A white eagle stands for Poland.

All these eagles of heraldry aptly convey the impression of strength and courageous defiance characteristic of the real eagle which inspired them.

Rough-legged hawk on nest (Alaska)

During April and again in October and November rough-legged hawks pass through our countryside. In size and build they resemble our common red-tailed hawk but their

Rough-legged Hawk

shanks, which are bare in most birds of prey, are feathered down to the toes, hence the name rough-legged. The rough-leg's tail, instead of being rust coloured as in the red-tail, is white with a broad dark band across the end. Rough-legs nest in the sub-arctic and arctic along the northern edge of forests and out on the treeless tundra. In their summer home they feed almost entirely on lemmings while in our part of the world they prey on mice and voles.

Many years ago I found a rough-leg nest on Bank's Island in our western Arctic. It was picturesquely situated on a bastion-like projection of a mud cliff at the foot of which a stream ran down to the sea. Beyond the stream was a gentler slope with a green cover of small tundra plants. Looking toward the mouth of the stream, one saw the shore lead of open water with small islands of ice floating in it. Beyond it the solid sea ice stretched away to the horizon. The nest was about three feet across and made of sticks of dwarf willow with dead grass in the center. As is characteristic of the buteo class of hawks some of the willow twigs in the nest still carried their green leaves. Another nest I saw in arctic Alaska was in a cleft on the side of a rounded rock about forty feet high. This rock and others sat like gigantic puffballs on a gently sloping hillside, probably deposited there long ago by glacial action.

The day after my neighbour had taken off a crop of oats in mid-October, a rough-legged hawk quartered the field all day. In the stubble it was easy for the hawk to spot its small prey. It would fly to and fro at no great height and sometimes hover over one spot. Rough-legs hover far more than any of our other large hawks. Occasionally the hovering would end in a plunge to the ground when the hawk spied a mouse or vole. Once or twice when the rough-leg flew along the woodland edge on one side of the field, a robin-sized grey bird darted out from the wood and mobbed the hawk for a few moments. This aggressive grey bird was a northern shrike.

The last rough-leg has generally moved on before the bald eagles have left Hastings Lake, but when the snowfall is scanty and there are plenty of mice, a few rough-legged hawks may be seen even in mid-winter and these probably stay until spring.

Northern shrikes like the one I had seen mobbing the rough-legged hawk, are winter visitors but in our area they are seen more often in the fall and again in early spring **Northern Shrike** than in the depth of winter. They are elegant birds, pale grey above and white below with a black mask running through the eye, much black on the wing and a black tail. Like all shrikes, and there are a number of species in other parts of the world, northern shrikes have a hooked bill like the hawks. They prey on small game like mice, small birds and, in summer, large insects. They do not have the strong talons of hawks but have feet like any of the song birds. Perhaps because of this they have developed the habit of impaling their prey on the barbs of wire fences, or on thorns, as a way of holding the prey while they tear at it with the beak. The predatory ways of shrikes account for the scientific name *lanius*, Latin for butcher. All shrikes were formerly called butcher birds in England. Even more extreme though not, I think, accurate is the German name for the shrikes which means throttler.

Northern shrikes breed in the forests of the Yukon, North West Territories, northern Manitoba, Quebec and Labrador. The only area in Alberta where they have ever been reported nesting is Lake Athabasca. I doubt whether they still do so since I never saw one during the time I spent in that area. In spite of their predatory habits shrikes do not endanger other wildlife populations. They are too scarce to have a significant effect on the numbers of their prey.

Snow geese overhead

They also have the great merit of being one of the very few of our birds which sing in winter. When there is still snow everywhere in early March a shrike's song delivered from a bush or small tree has a decidedly cheering effect. A shrike may begin his song by a few harsh calls but then he becomes quite musical and though he still interjects some grating notes among his sweeter sounds the overall effect is pleasant.

While the northern shrike is a winter visitor the rather similar loggerhead shrike is with us only during the summer months. It is a bird of open country and I have never seen one in the Hastings Lake area though it occurs in our general region.

Another fall singer is the tree sparrow. I recall waiting for ducks beside a small pond in late October. The foliage of the surrounding

Tree Sparrow willows and poplar was bright gold under a cloudless sky and every few minutes the sweet melodious song of one or more tree sparrows rang out. As it happened the ducks never came back but the sun shining on this isolated spot with its cheerful small birds made the time spent there well worthwhile.

In late September and through October flocks of some of our largest birds can be seen on their southward migration. There are

Snow Goose sandhill cranes and whistling swans flying in long lines and calling melodiously. These two have already been mentioned in connection with the spring

150

migration but snow geese must now be added. They are hardly to be seen in the district in spring but in the fall flocks heading south and flying at no great height are a regular feature. Until about 25 years ago, snow geese stopped over at Beaverhill Lake in great numbers every year. I had my first sight of these huge flocks in mid-April in 1949 with the late Professor Rowan, who was the first head of the Zoology Department at the University of Alberta and a keen bird scientist. We found a mass of at least 30,000 snow geese which covered half of the large slough adjoining the south shore of the lake. Other flocks were scattered over the rough pastures around the slough. There was a continuous din of the rough "kahak, kahak" calls of so many geese. As we approached, all of them, with a greater clamour than ever, took wing and for some moments darkened the sky ahead as they rose. The lake was still frozen save for a strip along the shore and the geese soon found a refuge on some sand and mud banks between this strip of water and the shore proper. They rested there until mid-afternoon. Then flock after flock began to lift off and fly in V's or in straggly lines to the south. We followed in the car and found the farm, 4 miles from the lake, where they had settled in a dense mass. Some were on summer fallow. Others were grazing on pasture. With them were a few Canada and some white-fronted geese. The owner of the farm told us the geese had been coming to his place for 10 days. They arrived around 6 in the morning, generally went back to the lake at mid-day and returned to stay till dusk. Presumably they would spend the night on the lake.

Sights of this sort are not to be seen at Beaverhill lake anymore. Though small numbers of snow geese can still be seen there in spring, the great flocks now follow a more eastern route along the Alberta-Saskatchewan boundary.

Thirty years ago I was able not only to watch snow geese on one of their main breeding grounds in the Canadian Arctic but to arrive there ahead of them. At the time five eskimo families living at Sachs Harbour were the only inhabitants of Banks Island. The head of one of the families, whom I'd met on the mainland while working as a medical officer for the Indian and Eskimo Health Service, was quite agreeable to being my host during my stay on the island. The land was still completely snow-covered and the sea ice was several feet thick when one of our neighbours, Bertram Pokiak, reported that he'd seen two flocks of snow geese heading landward over the sea ice on May 17. Next day little Navaluk, one of the children, came running into the house crying out "kanguit" -the geese. Indeed we could soon hear their harsh calls, a sound

Snow goose nest

which thrills every lover of nature when he hears it for the first time in spring.

Outside we could see the first small flock winging its way over the sea ice toward us, fine white birds with black wing tips standing out against the grey sky. On the southern slopes of the nearby coast hills there were a few snow-free patches and there I found some of the geese feeding on the short grasses next day. From then until mid-June flocks of snow geese appeared every day out over the sea ice flying toward the island. Some followed the shore northward but most flew inland through a gap in the coast hills heading directly for their breeding place forty miles north east of the settlement. At the peak of migration in late May up to two thousand geese passed over Sacks Harbour each day.

The eskimos made goose decoys of snow with a cap of moss to suggest the dark beak. When the birds were in sight they would lie on their faces and vigorously call in imitation of the geese. Again and again a passing flock would swing towards the hunters and come gliding into the wind towards the decoys. When they were overhead and in range shots would ring out.

In those days the Banks Islanders made a trip to the goose colony every spring to collect a load of eggs and I was one of that year's party. On the last day of May we set out with three sledges pulled by dog teams. As soon as we had passed through the coast range there was not a patch of snow-free ground to be seen. Some spring migrants had arrived. We saw a pair of rough-legged hawks and further on two cranes. Toward evening we could see the bluffs along the far side of the Egg River at a point some 26 miles from its mouth, which is the nesting area of the snow geese. Soon we were out on the flat valley floor. Here too the ground was generally snow covered but there were raised areas of bare ground separated by still frozen branches of the river. Thousands of geese stood on these snow-free patches and many had already laid eggs in spite of wintery conditions. Thousands more were standing about in masses on unbroken snow fields as far as the eye could see along the valley floor.

The goose nests consisted of a ring of dead grass on the mud surrounding one to five eggs. A variable amount of grey down which the birds pluck from their breasts lay around the eggs. These were all early nests. Had the geese been incubating for some time the eggs would have been lying inside substantial heaps of down. The female geese are fertilised before they arrive on the breeding grounds and are ready to lay their eggs immediately on arrival. For this reason we came across a number of eggs lying exposed on the mud without the birds having had time to make even a scrape.

We camped in the goose colony that night and next day my native companions busied themselves loading their sleds with goose eggs. I estimated that there were altogether about 30,000 snow geese and a few Brant geese in the colony. This was in 1953. Since then this colony has grown a great deal. There were over 200,000 birds there when a Canadian Wildlife Service scientist visited it in 1973.

This may have been due to a shift from another formerly important breeding colony on Wrangel Island which lies north of the eastern tip of Siberia. The Wrangel Island geese winter in North America, in the same areas as those from Banks Island. In 1960 there were about 200,000 snow geese on Wrangel but that number has gradually dwindled due to persistence of snow during the egg laying period.

Geese were not the only form of life we saw in the Egg River valley. A few barren ground caribou, the first I had ever seen of these interesting animals, were wandering about singly on the snow-covered slopes beyond the goose colony. There were jaegers of two species and a variety of shore birds which breed on the Arctic tundras. These included black-bellied and golden plovers, ruddy turnstones and pectoral, buff-breasted, semi-palmated and white-rumped sandpipers as well as sanderlings.

On the return trip a very cold wind came up and in order to make our lunch stop more comfortable my companions pitched one of their tents between two of the sleds. Once the primus was going, boiling some eggs and goose meat, the atmosphere in the tent was quite snug. As on the southward trip we noticed that there were just about as many flocks of snow geese flying away from the colony as there were others heading toward the breeding area. These must have found conditions at the Egg River too inhospitable.

Later that day an incident which could have been unpleasant but actually turned out to be merely amusing occurred. One of my host's sons aimed his .22 at a goose overhead. Being convinced no one could hit a flying bird with a rifle bullet I paid no attention. Suddenly I heard Pat, my companion, shout "watch out doc!". He pulled me aside as the goose came hurtling down, just missing my head.

Observations on banded birds have shown that the pair bond on geese generally lasts for life. I saw a striking example of the strong attachment between a snow goose and his or her mate on my return from the Egg River.

One day I spotted a goose lying dead on the ground some distance from the settlement. There had been no goose hunting since our trip and this bird must have been shot at least 17 days earlier. As I walked up to it another goose suddenly fluttered out of a depression in the ground and almost immediately landed again only a few paces away. Evidently it had been keeping a vigil beside its dead partner for over two weeks.

Later the same day I explored a little valley in the tundra with several small lakes. By now the coastal lowlands were completely snow free. On grassy inlets in one of those lakes glaucous gulls and black Brant had their nests. Further on I spotted the head of a snow goose above the marsh vegetation and found it had a nest with 5 eggs. They were warm to the touch and the goose had evidently been incubating.

In order to observe and photograph I set up a blind near this nest. When I returned to it for the first time the pair of geese were quite some distance from the nest so that I feared they had not accepted the blind. However, hoping for the best, I went into the blind. For a long time no geese could be seen through my peephole. After what seemed like an age there were 5 snow geese on a nearby pool. They came ashore and, feeding as they went, walked very slowly toward the nest. Suddenly the leading goose took wing and flew away. By now I was feeling very cramped and decided to wait just another half hour. Nothing happened during the next 15 minutes. Then a goose flew up and came down again considerably nearer the nest. Slow walking and a few more wing beats brought her to the nest and she was settled on it just before the end of the half hour. To my chagrin I found I had only one exposure left in the still camera but I had also brought along a movie camera and was able to shoot a few feet of film with that.

I returned to the blind on June 27 when both geese were together by the nest. Four of the eggs had hatched and there were the attractive downies, bright yellow with pale grey backs. The last egg had a hole in the shell and the youngster inside was piping in a lively manner. The parent geese stayed within a few feet of me but they started very slowly to lead the young toward the nearest pool. I thought it best to leave the scene so that they might not forsake the youngster that was about to hatch in the nest.

I had one more interesting encounter with snow geese during my stay on Banks Island. In mid-July the geese, having shed their large wing feathers, are unable to fly for two or three weeks. There were several hundred such flightless geese in the valley of the Kellet River a few miles from Sachs Harbour and the eskimos had the custom of driving some of these into their settlement to be killed for food. This and the annual collection of goose eggs at the Egg River are probably things of the past now that civilization is represented at Sachs Harbour by an RCMP post and a school. I went along on the goose drive. There were four of us, yet over a thousand geese marched in a dense flock ahead of us like a herd of sheep controlled by several sheep dogs might have done. The flock instinct of these geese is so strong that they make no attempt to scatter and one man alone can manoeuvre a flock in any direction he desires.

This goose drive yielded some results of scientific interest as well as meat for the eskimos. I had brought along some bands and placed one on every goose that dropped out of the march from exhaustion. A bird that could not keep going would raise its wings and start trembling, soon it would lie down and let the flock go on. After a rest it would of course recover.

Of the 78 geese I banded 9 were shot in winter; most of them in California, two in Oregon and one in Mexico. Five were shot during their fall migration, four of them in Alberta and two in Saskatchewan. One of the birds gave some indication of how long a snow goose can live if it's lucky enough to escape hunters. It was recovered 10 or more years after banding.

Winter Scene

Winter

Fred Rourke lives in a house which stands on a rise above the north shore of Hastings Lake. Through his kitchen window he can look out over a large part of the lake and its islands. He tells me that when he has seen the last eagle on the lake at the beginning of winter he is always overcome by a feeling of sadness. The departure of the eagles means that during the coming months he will see only those few birds that stay with us the year round and maybe one or two winter visitors. Life outdoors is thus sadly impoverished.

About 290 kinds of birds can be seen in Alberta, but only 40 of them stay through the winter. Some of these, like the white-tailed ptarmigan of the mountains or the spruce grouse of the north woods, need special habitats not found in our district. At best 25 kinds of birds can be expected there in winter. Some of these are rare or secretive, and an average winter walk is likely to yield sighting of only about six species.

The landscape of early winter, before the snow but when the waters are already frozen, has a charm of its own. The woods look dull and lifeless, though a splash of colour **Ruffed Grouse** is added here and there by the red of a stand of willows. However, on a sunny day the dead grasses, stubble and undergrowth in their various shades from brown through ochre to gold, can look bright and cheerful. It is at this season more than any other that I seem to see domestic cats out mousing near every farm. For long periods one walks in complete silence, until sooner or later the gay twittering of a group of black-capped chickadees is heard. Suddenly and with startling effect, a ruffed grouse whirrs up and away. The plumage of these birds blends so well with the dead leaves of the woodland floor, one rarely notices them before their sudden takeoff. Yet there are times when they give themselves away by a soft "petee" call, uttered while they step to and fro apparently unable to make up their minds whether to fly up or stay. Occasionally the sharp "keek" call of a downy or a hairy woodpecker rings out and a blue jay or a magpie may show itself. On lake shores, the general silence will be broken sooner or later by cracks, booms and other mysterious sounds coming from the ice.

For a day or two a frost may festoon every twig with glittering miniature icicles and carpet the ground with glistening velvet, but sooner or later the snow will come.

Under an overcast sky, the snow-covered landscape looks dull and melancholy, but everything is changed when the sun shines on glittering snowfields and creates blue shadows.

An uncommon but characteristic winter sight is a large grey hawk, generally seen making its way from one woodlot to another

Goshawk
or flying low over the tree tops. This is the goshawk, built, like the considerably smaller Cooper's hawk, with a long tail and short rounded wings. It is a great bird killer, particularly of ruffed grouse, but it is equally eager to take snowshoe hares. The Cree Indians call it "peeponassoo", from "peepon" meaning winter. Winterhawk is a good name for the goshawk. Apart from the tiny merlin and the gyrfalcon, a rare winter visitor, it is the only hawk that winters here. The second word of its scientific name *Accipiter gentilis*, harks back to the days of medieval falconry when only nobles were allowed to hunt with a goshawk. It would be released from the falconer's wrist when potential prey was flushed by beaters or dogs. In a short, swift flight, it would then bring down game birds or the large European hare.

The Hastings Lake area does not seem to suit the merlin. Perhaps it is too wooded for its requirements. But there are merlins

Merlin
in the city, where about ten pairs breed. From my office on the campus of the Uni-

versity of Alberta I see one almost daily between October and March. It perches on a TV antenna on the roof of the Education Building. No doubt it rests there between hunting flights, for these little falcons are skillful at capturing waxwings and house sparrows. Merlins were formerly called pigeon hawks. Watching pigeons fly or walk past the hawk on the Education Building on countless occasions, neither party taking the least notice of the other, the old name seems inappropriate. Merlins almost always prey on small birds, but there are records of rare instances where one has captured a bird as large as a ptarmigan or a pigeon.

On the route from the city to our acreage there is a large slough where I once had an interesting encounter with a gyrfalcon. It hap-

Gyrfalcon

pened in November when the slough was frozen except for a sizeable pool of water near a point of the shore. The open water had attracted about 50 mallards, of which I hoped to get one or two. As I walked toward the shore in search of some cover, the ducks naturally took off but as there was little open water elsewhere I felt certain they would be back. Sure enough, in about half an hour they came flying in over the ice. Strangely, while the flock was heading west, two birds were flying fast in the opposite direction. Binoculars identified these two as a grey gyrfalcon in hot pursuit of a drake mallard. When the duck came over the east shore of the slough, perhaps realizing it could not keep ahead of the falcon, it came down on the shoreline. This saved it. The falcon landed about 15 yards from the duck and made no attempt to go nearer. It simply stood there, soon attended by three magpies, as if it did not know what to do next. That may well have been the true situation. Falcons almost always strike their prey while in flight. Gyrfalcons will sometimes kill ptarmigan on the ground, but this particular falcon, perhaps a bird of the year, seemingly had no experience of this. In any case, I did not give him much time to make up his mind. My approach for a closer look caused him to fly away. However, he did not go far, because no sooner had I returned to the edge of the open water when he came speeding along, this time chasing a hen mallard. She pitched down on the water in front of me and the falcon climbed up above her and for a moment hung in the air. I think he was about to swoop down on the duck, but just then became aware of me standing there, for he swung off in a curve and was soon lost to sight.

My own hunting that morning was only slightly more successful than that of the falcon, of which I saw no more. After a while, two ducks appeared and set their wings to glide down on the water. I shot one of them, but it turned out to be the smallest, thinnest, most miserable little yearling mallard I have ever seen.

Out by our cottage there is a stand of chokecherries occasionally visited in winter by a few evening grosbeaks. What these striking

Evening Grosbeak yellow, black and white finches find there at that time of year is a mystery. The berries are thoroughly harvested in late summer and fall by robins, cedar waxwings and warblers. About Hastings Lake these grosbeaks show up around houses, where they are fed sunflower seeds, in the same large, noisy flocks that are such a feature about the feeders in city gardens after Christmas.

Another small bird of similar size which is also more often seen in the city than in the country is the pine grosbeak. The young

Pine Grosbeak males and females are rather drab, but the adult males make a beautiful sight, for they are bright red with black, white-barred wings and a dark tail. Often their cheerful, melodious whistles are heard before the birds themselves are seen. Their gentle calls always remind me of my first weeks in Canada, when in a November many years ago I first came across these colourful birds on the river bank in what were then the outskirts of the city.

Almost anywhere in open country, on stubble or near stacks of grain, one may come across flocks of snow buntings. While on the

Snow Bunting ground, these birds look like pale brown sparrows, but when they fly up a flash of white wings distinguishes them at once from all our other small birds. They come from the tundras of the Arctic. There I have often watched the males in their elegant black and white breeding plumage deliver their pleasant tinkling song from a rock or while on the wing. They nest in holes, usually in some cranny of a cliff or in a rock pile. However, at Chesterfield Inlet on Hudson Bay, where I spent a summer, there were snow buntings but hardly any rocks. There the buntings nested in the corners of disused buildings and in rusty tin cans lying on the ground. I was also struck by the way the birds concentrated on the few snowfields left in midsummer. They were after insects which had been immobilized by the cold and were particularly conspicuous on a background of snow.

Snow buntings breed in some Arctic habitats which almost no other birds visit. Most of Greenland is covered by a massive ice cap but in places the tops of mountains project through the ice, forming rocky islands called "nunataks". The only birds which live on these are snow buntings and rock ptarmigan. While "nunataks" are usually found in the outer zone of the ice cap, a large one

*Snow bunting in
winter plumage*

where these two birds also nest lies 40 miles in from the edge of the ice and is therefore also that far from the ice-free land of coastal Greenland.

On February nights, the "hoo hoo" calls of great horned owls are heard more than any other time. Though it is still the depth of

Great Horned Owl

winter, these hoots are a sign of spring. They indicate that the owls are already pairing up and marking out their breeding territories. In our area no other bird nests as early as they do, for they will have laid their full clutch of two or three eggs by late February or early March. Like most of our owls they do not build a nest, but take over an old hawk's nest, most often that of a red-tailed hawk. Wandering through the snow beneath the poplars, one may spot the owl's large head above one of these clumps of sticks. With its ear tufts, the owl appears cat-like as it looks down, but binoculars, revealing the large yellow eyes, soon dispel the resemblance. The owl seems perfectly aware of its safety on its elevated perch and after one look down at the observer returns its head to the resting position and resumes its calm gaze into infinity.

Outside of the nesting season, the great horned owl spends the daytime hours in hiding, perching close up against a tree trunk and under cover. They show up in the open in the early morning hours and at dusk, and may then perch quite prominently on a tree top. Usually large birds are much scarcer than small ones, yet the great

Hawk owl

horned, one of our largest, is by far the most common of our owls. On the other hand, its close relative in Europe, the eagle owl, is a very rare bird which has vanished from many of its former haunts.

The short-eared is the only one of our owls which generally migrates elsewhere for the winter. The medium-sized long-eared and two small species, the saw-whet and the boreal owl, stay in our woods all year round, but I have never encountered any of them in winter.

I have, however, come across a hawk owl. This 14-inch bird nests across the coniferous forests of northern and western Alberta

Hawk Owl

but appears in the parklands as a scarce winter visitor. It is easily spotted for, unlike most owls, it is quite active in full daylight and chooses well-exposed perches from which it watches for mice and voles. It nearly always tolerates a fairly close approach. With its barred underside and long tail, it has something of the smaller hawks about it, but its large rounded head with big eyes set in a flat facial disk is clearly that of an owl. When its wait for the sight of a small rodent has been unrewarded for too long, it drops down head-first from its perch, levels out, flies swiftly to another tree, and lands there from a steep upward glide to watch from the new lookout.

Occasionally a small herd of deer is seen far off across a snow-covered field against a backdrop of woodland. Or, closer at hand,

White-tailed Deer

a group may cross a secondary road. At this time of the year, the deer are grey and their winter coats certainly blend with their surroundings in a way which their rich reddish-brown of summer would not. In winter deer are generally in groups. The fawns born in the preceding June keep with their mother and even her offspring of the year before generally rejoin her for the winter. In this way, a group of seven or eight forms about a doe who is mother to them all. Though the rutting season of white-tailed deer runs from October through early winter, contact between the sexes must be brief, for I have only rarely seen a buck with these winter groups.

Apart from some of the bats, none of our mammals migrate, but they too, like the birds, are less evident in winter. Ground squirrels and badgers have gone underground to hibernate sometime in October and skunks go into their winter sleep toward Christmas. Muskrats remain active but stay in their houses or the push-ups of marsh vegetation that one sees on the ice of almost every slough. They get to their push-ups by swimming under the ice and rest there while eating water plants they have brought up from below.

Beavers have also vanished from the scene. They are either in their lodges or swimming under the ice to and from the store of

Beaver

poplar branches they have submerged nearby in the fall. At a glance, a beaver lodge looks a mere disorderly jumble of large and small sticks and branches, but if one comes upon one in the fall, one can see that most of it is plastered with mud. It is covered with more sticks and branches later on. The mud keeps out the cold, while ventilation is possible through the very top of the lodge where the roof is thinnest and free of mud. When mud and sticks freeze together the walls become so strong that even powerful predators like bears or wolves cannot tear them apart. Inside the lodge, the beavers live comfortably on a dry floor from which plunge holes lead down under the ice.

An animal that is as likely to be seen in winter as at any other time is the porcupine. Clinging in a young poplar or a willow, it

Porcupine

will be chewing away at the bark. If it is already high enough to be out of reach, it shows no alarm when an observer comes up close. It merely stares down with a vacant absent-minded look on its face and then returns to its business as if no one was there. Should it be met at a lower level, it climbs up with movements so painfully slow that I am always reminded of a sloth. A porcupine is just as slow on the ground, and this has occasionally saved people from starvation

Porcupine
Photo by William Rowan

when they are lost in the woods. The animal cannot escape a man armed with no more than a stout stick and it is of course quite edible. A porcupine is covered, except on most of the head, not only by spines but also by coarse hairs which are even longer. Hairs and spines hide most of its structure, so that when clinging to a tree it looks rather like a rough-surfaced sack with a small head.

In case I have exaggerated the porcupine's slow motion I must add that when I have seen one walking across a road it showed itself capable of much the same speed as a small dog. When pursued on the ground porcupines can accelerate to a clumsy gallop but even then they are easily overtaken. When approached in the bush they naturally make for the nearest tree and start climbing. If they cannot reach one in time they try to get their head under the protection of some log and raise their quills as they present their backs to the would-be attacker. Then they swiftly lash the spiny tail from side to side. The spines are only loosely attached to the skin and near the tip they bear scaly projections which act like the barbs on a fish hook so that they readily get stuck in the attacker's skin. As every countryman knows this is what happens when a dog attacks a porcupine. The dog returns whining to its master who, with a pair of pliers, has to remove the spines one by one. They will not come out on their own and are merely driven deeper by the dog's attempts to rub them off. After a similar experience wild predators generally die a lingering death for the embedded spines interfere with chewing and swallowing and the punctures are liable to get infected.

Yet there are predators which know how to dine off porcupines. The specialists at this game are the fisher, bobcat and wolverine. The evidence of tracks in the snow and the remains of the porcupine indicate that they attack porcupines in the trees. Slashing at a porcupine from below the predator knocks it out of the tree, then somehow rolls the fallen "porky" onto its back and attacks its undefended underparts.

In spite of this and the fact that they give birth to only one young at a time, these large rodents are still common. As many as 28 per square mile were found on a survey in Maine.

One wonders how these prickly animals mate. Their courtship and mating, which takes place in November and December, has been observed. Males follow females with grunting and humming sounds which can be heard for quite a distance. Only after some time does the female allow the proceedings to go further and somewhat comical "games" ensue. Both animals may rise up on their hindlegs and walk toward one another grunting or whining. They may rub noses, take a swipe at the partner's head with a paw or roll him or her over. Only after the two animals have been together for some time is the female ready for mating. It may be assumed that she is then so used to her mate that she will keep her spines lying flat while he sets about his task.

The porcupine's feeding on the bark of trees of several species makes it perhaps the most serious forestry pest among mammals. Surprising as it seems in so clumsy an animal, porcupines are reported to swim quite well, floating high in the water. Probably their hollow air filled quills give extra bouyancy. The purpose of their swimming is most often to get water lily leaves, evidently one of their favorite summer foods.

Before coloured glass beads became available, Indian women of many tribes embroidered leather clothing and moccasins with short segments of porcupine quills. Using white and dark parts of the spines, pleasing patterns could be created. Some tribes also made headpieces, which were worn like a crest, from pieces of porcupine skin with its hair and quills. Embroidery with porcupine quills was so highly developed that an entire book has been devoted to the many styles and variants of this craft.

Mice, voles and shrews are all quite active through the winter but are seen less often, for they do much of their moving about under the snow. Thinking of these small **Masked Shrew** animals brings to mind the drama revealed

Long-tailed weasel looking out of gopher hole

in the oat bin in the porch of our cottage. Because the oats were in a plastic bin with a lid that did not fit quite rightly, small rodents could get in. However, the smooth sides of the bin made it impossible to get out again. When I opened it one day, there was the slightly shrivelled corpse of a masked shrew, the remainder of another one, reduced to skin and bones, and the skin and skeleton of a red-backed vole. Masked shrews are quite common in our area. Like all shrews, they are fierce little carnivores, and though they generally live on insects and worms, they are quite capable of overcoming and killing small rodents larger than themselves, like the vole in question. Reconstructing events, the shrews first killed and ate the vole, for unlike the vole they could not eat the oats. When they became hungry once more, the stronger shrew killed and devoured the other one. After that, since no meat of any kind was left in its prison, the cannibal shrew starved to death.

The human observer may not see much of the small rodents in winter, but their arch enemies, the weasels, know how to find

Long-tailed Weasel

them at any time of the year. Our three species of weasels can be graded according to size, and in our district their prevalence goes according to their size. The largest, the long-tailed weasel about 10 inches long, is the one most often seen. I once stopped the car to get a good look at one of these graceful animals as it was moving over the snow-covered rough grass beside the road carrying a vole in its mouth. A magpie was hopping about the weasel, evidently hoping it could scare it into dropping the prey. It was unsuccessful and as I came closer the magpie took off, but soon the weasel too became alarmed. It dropped its dead vole and scurried up a pole that was leaning against a fence post. I came no closer and the weasel very soon regained its confidence, for it crept down the pole, seized its prey once more and took off with it.

The somewhat smaller short-tailed weasel must be scarce about Hastings Lake, for the only ones I have seen were two trapped a few years apart on our acreage. The least

Least Weasel

weasel, the smallest of them all, is also the rarest. The author of "The Mammals of Alberta" took only two of these during a winter in which he trapped 150 of the other two kinds. I have only seen these little animals in winter, on both occasions under the same circumstances. What looked like a wind-blown wisp of half-frozen snow appeared to be wafting across the road in front of the car, but a closer view revealed that it had a head and tail. Being a least weasel, it was white all over. The other two species retain their black tails, even in their otherwise white winter coats. In summer, all the weasels have brown upper parts.

This seasonal change from a brown or grey summer to a white winter coat is also shown by the snowshoe hare, by jackrabbits and among other animals by the Arctic fox. Experimental research has shown that the coat colour in animals of this type is controlled by day length. When they are exposed to long days of artificial light in midwinter, they soon moult into a coloured coat. These conditions imitate the natural lengthening of the days which they experience in spring. Similarly, exposure to short days causes the animals to grow a white winter coat out of season. Long days stimulate the outpouring of a hormone which acts on the hair follicles in such a way that they form pigmented hairs; that at least is how they have been shown to bring about their effect in short-tailed weasels. These seasonal colour changes have no doubt evolved because a brown summer coat blends fairly well into the background of these animals at that time of the year but would make them very conspicuous on the snow. However, since short days bring on the change into the white winter coat, the animals develop this coat about the same time every year, regardless of whether there is a snow cover or not. When the snow is late in coming, snowshoe rabbits that have already turned white are very conspicuous on the brown fallen leaves of the forest floor and much easier to spot than at any other time of the year. In this way, control by day length sometimes fails in the short-term to adapt the colour of the animal to that of its background, but in the long-term there is little doubt that it benefits the animal concerned. It may well be the best control system possible. Temperature fluctuates far too erratically to be a good predictor of snow and if the animals were somehow activated by light reflected from the snow, there would be quite a delay until the change of coat colour could be brought about.

Red squirrels and the rarer, much more secretive, flying squirrels, both remain active throughout the winter. The red squirrel,

Flying Squirrel that common, confident and noisy little animal, has never interested me very much. The dainty flying squirrel is another matter. Many years ago, a fellow naturalist told me that tapping a tree which had a hole further up the trunk would sometimes cause a saw-whet owl to look out. The first time I remembered to try this maneouvre not an owl but a flying squirrel peeped out of the hole. Since then, I must have beaten the trunks of countless apparently suitable live and dead trees, but never since that time has "anybody" looked out.

Unlike other squirrels, the flying squirrel is nocturnal and it has the large eyes that are characteristic of animals that are active in poor light. These attractive little animals often visit the bird feeder of my friend, Fred Rourke, at night. Although they are reported to feed mainly on vegetable matter, including fruits and nuts, it is scraps of meat which bring them to Fred's feeder. The feeder is just outside his kitchen window but the light in the house does not bother the squirrels. On one occasion, Fred saw a great horned owl swoop down on one of the little animals — it managed to avoid the owl's talons in the nick of time, but very soon afterwards Fred heard a terrible squeal. Evidently the owl, in a second attempt, had caught the squirrel after all while it was on its way to its winter tree hole.

A cheerful "chit chit chit" heard from some small birds flying overhead or well up in a tree, most often a birch, is one of the

Redpoll characteristic sounds of winter. The birds are redpolls, small finches named after the red patch on their foreheads and crowns. Apart from this, they are brown above, white with brown streaks below and have a black throat. Males also have a red breast. Sometimes one first becomes aware of the presence of these birds when one notices a scattering of brown scales on the snow beneath the birches where the birds have been eating the seeds. They also eat the seeds of many weeds and will not despise grain. Most of these little finches will be common redpolls which have occasionally nested in the Edmonton area but normally only do so further north. The hoary redpoll is not seen as often as the common species, from which it differs by having an all-white unstreaked rump. The rump is streaked in common redpolls. The two species are not infrequently seen together but are so active that it is no easy matter to get a clear view of their rumps. The hoary redpoll is a more distinctly northern bird, nesting only in the subarctic and low Arctic. Through the winter, only the calls of these birds, which include a melodious "tooee", as well as the "chit chit", are heard but sometimes on fine days in March their lively song rings out.

The most handsome of our larger animals is undoubtedly the red fox. Not only does it have an elegant shape, it also has a strik-

Red Fox   ing colour pattern. Unfortunately, it is rare in the Hastings Lake area, where I have only once seen one of these animals. On a sunny day in January it was walking purposefully through the snow that covered the frozen lake, until it disappeared into the bush on the far shore. I believe foxes are scarce in this district because they favour fields or mixed prairie with brushy areas, and for them our part of the world is too heavily wooded. In the more open country to the east, as about Tofield, foxes are seen more often. There are, of course, red foxes (of another subspecies) in the coniferous forest of northern and western Alberta, but these do not, I believe, range as far south as our district.

The number of showshoe hares that live in a given area varies greatly from year to year. In the fall and winter of 1980-81 they

Snowshoe Hare   were common in the district. I remember seeing seven in an hour's walk. This fall, a year later, they have almost vanished and I have only come across one during the whole period between September and the end of the year. These changes in "rabbit" populations are well known; they involve a gradual build-up of numbers peaking fairly regularly every nine or ten years on the average. As mentioned in an earlier chapter on lynx, peak numbers are followed by very sharp declines, or a so-called crash. Further north, these cycles of abundance and scarcity are even more extreme, and "rabbit" numbers there may vary from just one animal per square mile to over 3,000.

The snowshoe hare is not the only animal that shows population cycles. Lemmings and some of the other small rodents build up to peak numbers and then crash every four years with considerable regularity. This happens in the Old World as well as in North America, but animal cycles of this type are restricted to the northern parts of the northern hemisphere.

Young snowshoe hare
Photo by William Rowan

The numbers of several kinds of predators, both mammals and birds, are strongly influenced by the changing population of hares or small rodents. During the times when these prey animals are common, the predators have more young than usual. In crash years they may not breed at all. Predators which depend largely on prey which undergoes cycles of abundance therefore themselves also show cyclical changes, predator peaks generally lagging somewhat behind those of the prey. In this way, the snowy owl population on the tundra reflects the lemming cycles.

Conclusion

I wonder about the future of the parkland. Can moose and lynx or the shy broad-winged hawk continue to live here when more and more of the poplar forest is broken up into acreages; when the network of paved roads becomes denser, when motor boats roar on the lakes in summer and ski-doos whine on land and ice in winter?

Most animals are adaptable or they would not have persisted in an area where these disturbing factors are already prominent. But their adaptability is limited and the outlook for much of the wildlife near large and growing cities is in some doubt.

Taking a wider view the situation looks more promising. There are vast areas where human pressure on wild animals is minimal and likely to remain so for generations to come. Furthermore several National Parks provide sanctuaries for many species. When the population of particular animals in the surrounding areas is much reduced, emigration from within the parks may help to restore it to normal.

Another (and very promising) feature is the growing interest among the general public in wildlife and conservation problems. Forty years ago corpses of hawks strung on fences were a common sight in the countryside about Edmonton. These sorry spectacles have long vanished not merely because it is illegal to shoot birds of prey but because farmers now know that they benefit from the rodent killing of our hawks.

Years ago a person might have heard something like the following in some country beer parlour: "Saw a big coyote coming up from the ferry"; and the inevitable reply: "Did ye get it?". Back then it was customary to have a twenty-two handy while driving the back roads so that business might be enlivened by a bit of "sport". I believe that this habit has very nearly disappeared.

The survival of wildlife depends on our management of this and other natural resources. If we can continue to develop this management, then the outlook is fairly promising. In all probability our descendants will also be able to enjoy the sort of scenes I have described.

Index

Date Due